A
Rational
Christian Look
at UFOs
and Extraterrestrials

# LIGHTS IN THE SKY
# &
# LITTLE GREEN MEN

HUGH ROSS ~ KENNETH SAMPLES ~ MARK CLARK

THE JOHN
ANKERBERG
SHOW

© 2024 by Reasons to Believe

Reprinted by The John Ankerberg Show with permission from Reasons to Believe / RTB Press.

All rights reserved. No part of this publication may be reproduced in any form without written permission from Reasons to Believe, 818 S. Oak Park Rd., Covina, CA 91724. reasons.org

Cover design by Dan Jamison
Cover photo by ©vcl/Spencer Rowell/GettyImages/FPG

Some of the anecdotal illustrations in this book are true to life and are included with the permission of the persons involved. All other illustrations are composites of real situations, and any resemblance to people living or dead is coincidental.

Unless otherwise identified, all Scripture quotations in this publication are taken from the HOLY BIBLE: NEW INTERNATIONAL VERSION® (NIV®). Copyright © 1973, 1978, 1984 by International Bible Society. Used by permission of Zondervan Publishing House. All rights reserved.

Ross, Hugh (Hugh Norman), 1945-
   Lights in the sky and little green men : a rational Christian look at UFOs and extraterrestrials / Hugh Ross, Kenneth Samples, and Mark Clark. p. cm.
Includes bibliographical references and indexes.
1. Occultism--Religious aspects--Christianity. 2. Unidentified flying objects--Religious aspects--Christianity. I. Samples, Kenneth R. II. Clark, Mark, 1955- III. Title.
  BR115.O3 R675 2002
  261.5'1--dc21

Printed in the United States of America


2 3 4 5 6 7 8 9 10 / 20 21 22 23 24

For more information about Reasons to Believe, contact (855) REASONS / (855) 732-7667 or visit reasons.org.

*For Joel and David;*
*Sarah, Jackie, and Michael;*
*Matthew and Caleb*

# Contents

ACKNOWLEDGMENTS
PREFACE

| | |
|---|---|
| 1. The UFO Craze    *Kenneth Samples* | 13 |
| 2. Types of UFOs    *Kenneth Samples* | 23 |
| 3. Life on Other Planets    *Hugh Ross* | 37 |
| 4. Evolution's Probabilities    *Hugh Ross* | 47 |
| 5. Interstellar Space Travel    *Hugh Ross* | 61 |
| 6. RUFOs—The Unexplained UFOs    *Hugh Ross* | 71 |
| 7. Government Cover-Ups    *Mark Clark* | 79 |
| 8. Government Conspiracies    *Mark Clark* | 91 |
| 9. Nature and Supernature    *Hugh Ross* | 105 |
| 10. The Interdimensional Hypothesis    *Hugh Ross* | 113 |
| 11. A Closer Look at RUFOs    *Hugh Ross* | 123 |
| 12. Abductees    *Kenneth Samples* | 135 |
| 13. Contactees    *Kenneth Samples* | 149 |
| 14. UFO Cults    *Kenneth Samples* | 159 |
| 15. The Bible and UFOs    *Hugh Ross* and *Kenneth Samples* | 173 |
| 16. Summary    *Hugh Ross* | 183 |
| APPENDIX A: Fine-Tuning for Life on Earth | 187 |
| APPENDIX B: Probabilities for Life on Earth | 203 |
| APPENDIX C: Fine-Tuning for Life in the Universe | 209 |
| NOTES | 211 |
| BIBLIOGRAPHY | 257 |
| NAME INDEX | 263 |
| SUBJECT INDEX | 269 |
| ABOUT THE AUTHORS | 277 |
| ABOUT REASONS TO BELIEVE | 279 |

# Acknowledgments

No book results from the authors' efforts alone. At times, our wives and children may have thought we were "abducted" by the UFO project. So thank you, Mara Clark, Joan Samples, and Kathy Ross. And thank you, kids—we dedicate this book to you.

Many members of the Reasons to Believe staff, including volunteer staff, also helped us in one way or another, finding reference material, making phone calls, asking tough questions, and editing drafts. Those who invested hours upon hours include Robert M. Bowman Jr., Bonita Connoley, Patti Townley-Covert, Jody Donaldson, Sandra Dimas, Marj Harman, Amy Jung, Linda Kloth, Kathy Ross, and Tani Trost. This book's authors *and* readers owe you a huge debt of gratitude.

Our friends at NavPress deserve thank yous for encouraging us to tackle a project involving daunting challenges, both intellectual and spiritual. Nanci McAlister, Lori Mitchell, Greg Clouse, Pat Miller, and Eric Stanford, who worked closely with us, deserve enormous respect and appreciation.

# Preface

SPECULATIONS ABOUT unidentified flying objects and extraterrestrial beings just won't go away. They continue to crop up in conversations all over the planet. Almost everyone can tell a story of seeing weird lights in the sky—lights that seem to defy explanation. Almost everyone wonders about "little green men" flying overhead in spaceships or at least about whether intelligent life exists somewhere beyond planet Earth.

"What are those things in the sky?"
"Where do they come from?"
"Does alien life exist?"
"If flying saucers are real, do we need to worry about them?"
"Is the government hiding something?"

Questions like these continue to be asked, revealing a weakness in the answers offered to date.

Often, spokespersons from fields such as astronomy, theology, philosophy, politics, and the military patronize those who pose questions about UFOs (unidentified flying objects), telling UFO observers that they really saw meteors, mirages, military aircraft, the planet Venus, or college students' clever pranks. And of course, in many cases, naturalistic explanations are the right ones. But the spokespersons' hastiness in offering such answers demonstrates the contempt they hold for anyone who is prepared to believe in UFO phenomena.

The experts' persistent denial of anything supernatural or paranormal drives many in the general populace to search elsewhere for satisfaction of their curiosity and concerns. Too often, though, the people to whom they turn are UFO advocates who play on emotions, short-circuiting scientific reasoning, good scholarship, and philosophical thought. These people have an agenda—often one involving the exercise of questionable moral and spiritual influence—and thus the curious may be drawn into a situation that is a threat to their well-being.

In short, people with questions about UFOs may encounter condescending rejection on the one side and dangerous credulity on the other. And it's because of such insufficient responses to legitimate questions that the three of us—an astronomer and Christian apologist, a philosopher and cult researcher, and a political scientist specializing in national security—wanted to write a book on this subject. Many people are intensely curious about unidentified flying objects and extraterrestrial intelligence (ETI) and are frustrated by the explanations they have received. For too long, the community of scientists and scholars who could and should be providing answers have avoided this topic. We wanted to fill the gap—and to do it from a perspective that reflects the Christian worldview we share.

As an astronomer who through the years has logged thousands of hours of observation time, I have learned that science can and does address the possibility of life's existence elsewhere in the universe. I began stargazing as a young boy, and by age seventeen I had become director of observations for the Royal Astronomical Society in Vancouver, Canada. Naturally, then, in my youth I began thinking about extraterrestrial life. Later, when I came to Caltech in the mid-1970s for postdoctoral research, the faculty assigned me the task of processing UFO reports. Now I was really focused on the subject of life elsewhere in the universe. And at about that same time, my belief in the reality of the supernatural led me to undertake an intense study of the Bible. I learned from the pages of Scripture about the all-powerful, always present, and ever-caring God of the universe and about His creation of life. The result of all this was that my curiosity regarding whether life could exist elsewhere and, if so, whether it could come to planet Earth was fully satisfied. Since then I've spoken to people all over the world about UFOs (among other topics), and many of them have told me that they, too, have found resolution and satisfaction in my discoveries.

These discoveries led me to write my chapters in this book. Chapters 3, 4, and 5 explore the conditions necessary for life to exist on other planets—and to show up in our atmosphere in spaceships. Chapter 6 then opens up the topic of residual UFOs, or RUFOs, which are those unexplainable yet real phenomena that remain after all naturalistic explanations have been exhausted. Further on, chapter 9 addresses the key question of the scientific evidence for the supernatural and of how nature and supernature intersect. That discussion leads naturally into chapter 10, which examines the hypothesis about UFO phenomena that I favor: the trans- or extra-dimensional hypothesis. Chapter 11 goes on to examine RUFOs in light of the trans- or extra-dimensional hypothesis. Finally, chapter 15 offers biblical insight regarding UFOs, especially RUFOs.

My chapters should answer many of the questions you have about UFO phenomena. However, while I can address questions about flying objects and extraterrestrial life from a scientific and theological basis, that information is not enough to cover the broad topic of UFO phenomena. Many other questions remain, requiring the expertise of my two colleagues: Kenneth Samples and Mark Clark.

Kenneth, a philosopher, teaches courses in logic and researches new religious movements, or cults. For many years he studied cults with the late Walter Martin, founder of the Christian Research Institute. Years of solid scholarship inform his expertise on alien abductions and UFO religions. Because thinking people deserve a logical rationale to supply the appropriate context for UFO experiences, Kenneth provides just that in his chapters. Giving an overview of UFO phenomena in chapters 1 and 2, he supplements his own extensive knowledge by citing world-renowned UFO experts. In chapters 12 through 14 he furnishes fascinating and profound insights into alien abduction, ongoing contact with aliens, and UFO cults.

Yet a discussion of UFOs and ETI cannot be considered comprehensive without addressing the possibility of government involvement in cover-ups and conspiracies. Mark Clark, a political science professor specializing in military and strategic studies, provides the breadth and depth of insight necessary to accomplish this task. In chapter 7 Mark exposes the three mysteries that are most often connected with government cover-up in people's minds: the supposed alien crash landing near Roswell, New Mexico; the government's UFO study known as Project Blue Book; and Area 51, where some think alien spacecraft and bodies are stored. Then in chapter 8 Mark offers a comprehensive look into conspiracy thinking.

Many years ago, a college teacher challenged the three of us to help people understand what lies behind UFO sightings and claims for ETI. For various reasons, other experts have been either afraid or unwilling to deliver the answers. Yet over the past few decades, frustration on the subjects of UFOs and ETI has continued to rise rather than to decline. The motivation for this book is the need to communicate clear, satisfying explanations from scientific, theological, philosophical, and political standpoints. By tackling this problem from a variety of disciplines and with a holistic approach, taking seriously the revealed truth from God contained in the Bible, we, the authors, intend to answer the legitimate questions connected with UFOs and ETI.

We hope this book will compel people to explore beyond surface explanations. We want to encourage rational thinking, sound logic, and critical evaluation. The nature of the subject demands these tools, and we believe that our treatment of the topics of lights in the sky and little green men puts them to good use.

*Hugh Ross*

CHAPTER 1

# THE UFO CRAZE

*Kenneth Samples*

Hector has a special place he goes to when he needs to do some undistracted thinking. He drives his pickup out into the California desert and down a rutted road for miles until he arrives at a west-facing cliff. There he sits with his legs hanging over the brink and watches as the sun sets in a spectacle of reds and oranges. He continues to sit as the sky turns to blue and then to black and the stars come out in a profusion he never sees from his backyard in the city.

One such night in the desert he saw an unusual light moving rapidly across the sky. At first he thought it was a plane or helicopter. But it was moving too fast. And then it made a sudden, sharp turn that no ordinary aircraft could execute. Hector got to his feet, his every sense alert. The object came nearer until it slowed and hovered over the ground less than a quarter mile away. Hector was trying to decide whether he should flee or stay when an intense beam of light suddenly shone from the base of the craft onto the ground. Hector knew something big was about to happen.

FROM ANTIQUITY, individuals have reported seeing unusual and inexplicable things in the skies. Often people observed real objects—natural phenomena that only later could be understood and appreciated in light of advancements in science, particularly in physics and astronomy. To those unfamiliar with astronomical or atmospheric phenomena, the ordinary can appear extraordinary. Nonetheless, some people insist that extraordinary flying anomalies have persisted throughout the ages.[1] Even today, some reports of strange sightings are difficult to dismiss as being misidentified natural phenomena, though natural explanations may yet be found for at least some of them.

Flying entity reports, whether subjective or objective, often come in waves, and some of these waves began rolling in just prior to the age of human aviation. For example, in the latter part of the nineteenth century, so-called "airships" were reported in the skies above the United States.[2] And during the early decades of aviation, reports of unidentified aerial objects began coming in from both commercial and military pilots. During World War II, Allied and Axis pilots reported observing mysterious aerial anomalies that paced their aircraft during flight. Both sides speculated that these "Foo Fighters," as they were called, were advanced enemy aircraft. No universally accepted explanation has ever been found for what these pilots reported.

Unidentified flying objects (UFOs) and the alien beings associated with them, known as extraterrestrials, as well as the possibility of life someplace besides Earth (extraterrestrial intelligence, or ETI), all continue to be sources of speculation. Is the human race alone in the universe? Or is there reason to believe intelligent life exists somewhere else in the cosmos? Could extraterrestrial life visit the Earth, and if it did, what would that encounter be like? These are some of the questions addressed in this book. And these were some of the questions that ushered in the flying saucer age, which began immediately after World War II and the appearance of the Foo Fighters.

## A Flying Saucer History

AS A NUMBER of UFO researchers have pointed out, the second half of the twentieth century has been called the "age of the flying saucer."[3] More precisely, the beginning of that age has been identified as three o'clock in the afternoon on June 24, 1947. That's the moment when private pilot and businessman Kenneth Arnold, while flying his plane near Mount Rainier in Washington, first observed nine bright objects traveling at incredible speeds for the time (estimated in excess of sixteen hundred miles per hour).[4] Arnold, a seemingly credible witness, described the objects as boomerang-like and disk-shaped, and he described their movement as appearing "like a saucer would if you skipped it across the water."[5] The headline of an Associated Press story mentioned "nine bright saucer-like objects," and the age of the flying saucer had officially begun, at least for the media. Arnold was not the only one to report observing those mysterious objects that day. At least twenty other reports of similar sightings, the majority originating from the Pacific Northwest, added credence to Arnold's story.[6]

Reports continued in the late 1940s, but the 1950s brought about a UFO obsession in America. Researcher Jerome Clark reports:

> The first books with "flying saucers" in their titles saw print in 1950. Speculation about UFOs became a national craze, with popular opinion divided between those who dismissed the phenomenon entirely (as the product of hysteria, hoaxes, and misperceptions) and those who saw it as of enormous potential significance. Some individuals became consumed with UFOs, and by the early 1950s the first UFO organizations were formed.[7]

Soon, for national security reasons, the United States government became interested in the emerging flying saucer phenomenon. The Cold War had begun, and the American military was concerned that these "ships" might be advanced Soviet aircraft. The military used a more conventional term to describe these aerial anomalies—"unidentified flying objects."

From the late 1940s through the 1960s, the United States Air Force investigated UFO reports through various projects and committees, including Project Sign (1947), Project Grudge (1949), and Project Blue Book (1951–1969). Just prior to the Air Force's closing of its investigation into UFOs, the Condon Report reflected the military's official conclusions on the subject. The report offered the following statement:

> The emphasis of this study has been on attempting to

learn from UFO reports anything that could be considered as adding to scientific knowledge. Our general conclusion is that nothing has come from the study of UFOs in the past 21 years that has added to scientific knowledge. Careful consideration of the record as it is available to us leads us to conclude that further extensive study of UFOs probably cannot be justified in the expectation that science will be advanced thereby.[8]

The lack of information forthcoming from the government, as evidenced in the Condon Report, gave birth to belief in a government cover-up about UFOs.

From the mid-1960s through the 1980s, UFO phenomena changed from sightings of possible extraterrestrial spaceships to associating UFOs and extraterrestrials with traditional paranormal and occult phenomena (ghosts, poltergeists, monsters, and so on).[9] Claims of visitations from extraterrestrial or supernatural beings became commonplace among UFO reports, with the abduction phenomenon taking center stage. In the last twenty-five years more books have been written on alien abductions of human beings than on all other UFO-related topics combined.

The 1990s contained their share of unusual UFO-related events. The possibility that alien spacecraft had crashed on Earth, and that the spacecraft and even alien bodies had been retrieved, remained a viable option for people who continued to believe that UFOs were visitors from interstellar space. The 1990s also saw a UFO cult become deadly for the first time. In 1997 thirty-nine members of the Heaven's Gate cult committed suicide en masse. They willingly took their own lives, believing they would find eternal salvation by joining the "mother ship," a spacecraft they believed was trailing the Hale-Bopp comet.

By their very nature, physical UFOs piloted by alien beings would validate the case for extraterrestrial intelligence if they were to show themselves to the world in an undeniable way. That sort of evidence remains lacking after more than fifty years of the flying saucer age. Yet UFO phenomena of different types continue to be reported.

## UFO Phenomena

WHILE THE LETTERS UFO stand for "unidentified flying object" and thus reveal the term's general or basic definition, the subject is so complex that it is increasingly difficult to provide an unambiguous definition for the term UFO. In fact, many serious UFO researchers now use the term "UFO phenomena" or "UFO phenomenon" instead of "UFOs."[10]

This nomenclature distinction, while it may seem trivial, suggests the difficulty the experts have in getting a handle on the subject.

Six difficulties arise when one attempts to provide an adequate and positive definition for an individual UFO, and these intensify when it comes to providing an adequate explanation for the phenomena taken as a whole. While these criticisms are not fatal to thinking about UFOs, they deserve careful reflection. Some may view these criticisms as excessively skeptical—but then the subject could use a little more commonsense skepticism! The six points overlap to some degree.

First, it has become increasingly difficult to distinguish between UFOs and their accompanying multifaceted phenomena. The two seem hopelessly intertwined, and therefore an adequate definition needs a combined logical, scientific, sociological, psychological, and religious assessment.

Second, in attempting to define a UFO, one is trying to identify that which is yet unidentified. This is similar to the problem in logic of defining a negative. It is difficult to provide meaningful definitions, not to mention classifications and categories, for phenomena that reside largely in the realm of the unknown. There is a certain logical legitimacy in reasoning from what is known to what is unknown, but beginning with the unknown and moving to the known is fraught with difficulties.

Third, it is impossible to perform a direct study of UFOs. Rather, UFO researchers study human *reports* about UFOs. One cannot overstate the difficulty of performing objective, scientific, and logical analyses of phenomena that come to researchers secondhand and mediated through intense subjective experience.

Fourth, UFO phenomena often involve bizarre occurrences. A report of these strange events often violates any previous definition that has been set down. Thus, subsequent UFO-related reports require attempts to form a new definition.

Fifth, definitions for UFOs suffer from both vagueness and ambiguity. Definitions lacking in clarity result because UFOs by their very nature defy precision.

Sixth, the meaning of terms changes over time. Originally, UFO simply stood for "unidentified flying object," but with the advent of UFO movies, saucer clubs, and unusual reports, UFO has come to mean "a spaceship with extraterrestrial life forms on board."

Despite these difficulties, ufologists (scientific or technically oriented researchers of UFO phenomena) and others still find it possible to speak intelligibly about UFOs. As long as one keeps in mind that UFO phenomena are diverse and complex, it is worthwhile to pursue a study of them.

## People Interested in UFOs

THOSE WHO SPEND a considerable portion of their time dealing with UFOs come from all walks of life, all socioeconomic levels, and all geographic locations. From hobbyists to scientists, farmers to astronauts, all types of people from all over the world are captivated by UFO phenomena.

Nine groups of people interested in UFOs are listed here, along with their basic conclusions concerning UFO phenomena.[11] The views ascribed to each group are not representative of each and every person within a given category but instead represent a paradigm, or model, for that category.

*Natural scientists.* By and large, the Western scientific community remains highly skeptical concerning the reality of UFOs, especially when these are understood as metallic craft transporting extraterrestrial visitors. Usually Western scientists maintain that UFOs are the result of misidentified natural or man-made phenomena. Many physicists, chemists, and astronomers, however, think that the existence of intelligent life elsewhere in the universe is feasible in light of natural evolutionary processes. They have launched "exobiology," a new branch of biology that considers the possibility of life's origin and evolution on different planets. But for most scientists, the search for extraterrestrial intelligence is separate from the highly questionable phenomena of UFOs. Recently, however, a distinguished panel of scientists concluded that some UFO reports may indeed be worthy of scientific investigation.[12]

*U.S. government officials.* Military investigations conducted in the United States from the late 1940s to the late 1960s concluded that evidence for an extraterrestrial interpretation of UFOs was lacking and that the unexplained UFO reports were likely misidentified natural or man-made phenomena. They also concluded that UFO phenomena posed no threat to American national security. The federal government partially funds NASA's SETI (Search for Extraterrestrial Intelligence) program, but overall it doesn't seem interested in the subject of UFOs. Due to consistent accusations of cover-up by some conspiracy-oriented UFO researchers, it is important to note that there doesn't appear to be any clear and convincing evidence that the U.S. government is or has been involved in a significant cover-up or conspiracy concerning UFOs.[13] Some key UFO researchers have asserted, however, that the official government investigations into UFOs have been superficial and flawed and that the government may have presented disinformation on the subject.

*Professional ufologists.* These people are divided over how to interpret UFOs. Some remain skeptical as to whether there is an actual, objective, and intelligent stimulant to UFO reports. Even those who affirm UFOs as an objective and intelligent reality are divided as to whether they are best explained as extraterrestrial visitors or as interdimensional (from a different dimension of reality) phenomena or even possibly as something else. The extraterrestrial hypothesis is popular among American ufologists. However, many leading ufologists in other parts of the world lean toward some form of the interdimensional hypothesis. Well-known UFO research organizations include the J. Allen Hynek Center for UFO Studies, the Mutual UFO Network, and the British UFO Research Association.[14]

*Social scientists.* Sociologists, psychiatrists, and psychologists generally begin with the presumption that UFOs are not an objective empirical reality (or any other kind of objective reality). Instead, most social scientists hold that UFO sightings and related phenomena are caused by social, cultural, or psychological factors.[15] Social scientists study the types of people who claim to have had UFO-related experiences, such as contact with extraterrestrials or abductions. Then social scientists propose natural explanations for these people's perceptions of the phenomena.

*Popular UFO enthusiasts.* For many people, UFOs are a hobby. They attend flying saucer conventions, form saucer clubs, and enjoy the flood of UFO-related movies. Popular UFO enthusiasts usually don't study UFO phenomena scientifically, nor do they view UFOs in a metaphysical or religious sense.

*UFO debunkers.* Some people's main interest in UFOs is to show the phenomena to be other than extraterrestrial or interdimensional. They do not believe that there are alien or interdimensional beings behind UFO reports, but instead they firmly believe that UFO phenomena can be explained naturally as the result of misidentified natural or man-made phenomena, human psychological experiences, hoaxes, and so on.

*New Agers.* New Age proponents and UFO contactees have a common connection—a religious one. Often those who claim to have been contacted by alien beings engage in various kinds of occult activities practiced by New Agers, such as channeling (spiritism), trance state, automatic writing, the use of crystals, poltergeist effects, psychic healing, and out-of-body travel.[16] Some New Agers believe they communicate with various alien beings through channeling. UFO religion is steeped in occultism, especially spiritism.

*UFO cult members.* Those involved in UFO cults generally believe that an advanced extraterrestrial civilization (from this solar system, from outer space, or from another dimension) is visiting Earth in order to save humanity from some form of catastrophe or to guide humanity in its continued evolutionary development. UFO cult members embrace an occult-based religion and look to UFOs and alien beings as their salvation "from above."

*Christian theologians and apologists.* Some evangelical Christian theologians are interested in the question of God's relationship to other worlds (exotheology). If intelligent beings exist elsewhere in the universe, what are the implications of this for Christian truth claims? In addition, Christian apologists are interested in UFO religion as a challenge to the truth claims of Christianity. While evangelical Christians debate the possibility of extraterrestrial intelligence, most Christian apologists view UFO phenomena as counterfeit religious phenomena with direct connection to occultism and probably also the demonic realm.

The nine groups listed here do not include the general populace—the millions who are fascinated with UFOs and the thought of extraterrestrial intelligence but are not among the avid UFO enthusiasts. The general populace may well be the group most served by the answers provided on the pages that follow.

CHAPTER 2

# TYPES OF UFOS

*Kenneth Samples*

The reporter from the newspaper was having a tough time compiling her story. The night before, more than a dozen calls had come in to the local police station telling of strange lights in the sky. But as she interviewed the callers, she could find little on which they agreed.

"There was one big, white light. Then it suddenly split into two lights that went off in different directions."

"It was a yellowish light going very fast in a straight line. I've decided it must have been a meteor."

"Something landed in the woods beside our house. Our dogs were going crazy. We locked our doors; there was no way we were going out there."

Sighing, the reporter began to despair of making sense of what had happened in the night.

IF UFOLOGISTS ARE going to have any hope of discovering what UFOs really are, then like any other researchers, they have to approach the relevant data in an organized way, with some possible categories and hypotheses already in mind. That's just what the best of them do. There are plenty of crackpot UFO enthusiasts out there, but many ufologists are respected scientists and other experts who are going about their task in a professional way. In the case of their particular discipline, though, the amount of data they are expected to sift through is vast.

One begins to see the magnitude of the case when the number of people who report having seen UFOs is considered. The Gallup Organization conducted polls of adult Americans in 1973, 1990, and 1996 regarding UFOs. Some 11 percent, 14 percent, and 12 percent, respectively, of those polled claimed to have personally witnessed a UFO.[1] Another polling organization, International Communications Research Services of Media, Pennsylvania, found in polls performed in 1990, 1991, and 1997 that 14 percent, 14 percent, and 13 percent, respectively, of American adults said they had seen a UFO.[2] Both polling organizations noted that about half of all Americans believe that UFOs are real and not just the result of people's imagination.

Given the vast number of people who claim to have seen UFOs, whether or not they report them, there is no precise way of knowing how many UFO sightings have taken place. But everyone agrees that it's a lot. J. Allen Hynek estimated during the 1970s that there were approximately one hundred UFO sightings each night throughout the world.[3] Researcher Jacques Vallée suggests that possibly only one in ten UFO sightings actually gets reported and that the number of sightings may in fact range into the millions.[4] Since the so-called "flying saucer age" began in 1947, thousands of UFO reports have been investigated by researchers.[5] The U.S. Air Force, under projects Sign, Grudge, and Blue Book, 1948–1969, investigated some fifteen thousand UFO reports. Still other thousands of UFO reports have been investigated by private UFO research groups, and ufologist Timothy Good says that over three thousand UFO reports have been generated from military and civilian pilots alone.[6]

Furthermore, UFO sightings are a worldwide phenomenon, with reports coming in from nearly every nation on the planet. In the United States, UFOs have been reported in every state, with the preponderance being in the northeastern and southwestern regions. The most common areas for UFO sightings are rural areas, small towns, and military installations. As UFO sightings come in waves, the greatest number of sightings have been recorded in 1947, 1952, 1966, and 1973.[7]

This chapter first explains how ufologists classify UFOs, then suggests natural explanations that can identify most unidentified flying objects, and finally introduces the two main hypotheses that are used to explain residual UFOs, or RUFOs.

## Systems of Classification

UFO PHENOMENA ENCOMPASS a broad variety of complex, unusual, and mysterious events that individuals claim to have witnessed or experienced. In fact, the phenomena at times seem so bizarre as to defy rational inquiry. UFO investigator John A. Saliba identifies four specific difficulties that make the study of UFO phenomena different from ordinary scientific inquiry:[8] First, UFO phenomena remain distant and elusive, and thus they "cannot be directly and thoroughly analyzed."[9] Second, UFO phenomena are difficult to categorize and there are "no undisputed experts who can verify the phenomena."[10] Third, because UFO phenomena often combine empirical data with religious and psychic experiences, scientists face difficulty in getting a handle on evidence. Fourth, UFO phenomena seem sufficiently mysterious to invite a variety of interpretations.

While there are real difficulties in studying UFO phenomena, two specialists, J. Allen Hynek and Jacques Vallée, set the standard for responsible UFO research. Each has come up with a useful system of classification for UFOs that is in regular use today.

J. Allen Hynek, former consultant to the Air Force's Project Blue Book and one of the cofounders of the Center for UFO Studies (CUFOS), set forth the first standard categories for UFOs in his groundbreaking book, *The UFO Experience* (1972). Virtually all scientists evaluating UFO phenomena use these categories.[11]

Hynek's categories encompass both distant sightings and close encounters. In the second group of UFO phenomena, however, Hynek's original categories ended at close encounters of the third kind (CE-3). When Hynek developed his system, three categories were adequate to cover the encounters that were being reported. But since the time Hynek set forth his categories, a flood of additional encounters have been reported that go well beyond CE-3. Some ufologists, therefore, now accept close encounter categories of the fourth and fifth kinds.[12]

Here's how Hynek's system looks with the augmentation of CE-4 and CE-5:

**Distant Sightings**
1. *Nocturnal lights* (NL): Phenomena observed at night, particularly unconventional lights.
2. *Daylight disks* (DD): Phenomena observed during the day, usually oval or disk-shaped objects in the sky.
3. *Radar-visual* (RV): Unexplained radar blips that coincide with visual sightings of UFOs.

**Close Encounters**
1. *Close encounter of the first kind* (CE-1): The UFO is sighted at close range (within five hundred feet) of the observer, but it does not affect or interact with the witness or the environment.
2. *Close encounter of the second kind* (CE-2): The UFO leaves physical effects on the environment (depressed or scorched ground, broken tree limbs, signs of radiation) and may even cause power failures (automobile engines stall, radios stop playing).
3. *Close encounter of the third kind* (CE-3): The living occupants of the craft are seen by witnesses; however, no communication or further contact takes place. This kind of encounter almost always happens at night and usually with only one or two observers.
4. *Close encounter of the fourth kind* (CE-4): Direct contact occurs between alien beings and the witness. Apparently taken aboard a spacecraft, the abducted witness interacts extensively with the spacecraft's occupants.
5. *Close encounter of the fifth kind* (CE-5): The observer suffers permanent physical injuries or death.

While Hynek's system of distant sightings and close encounters has proved useful, Jacques Vallée, physicist, computer technologist, and widely known ufologist, recently developed a new system for tracking and evaluating UFOs. Vallée's comprehensive twenty point classification includes three new standard categories: anomalies, flybys, and maneuvers, along with the standard close encounters classifications.[13] Each of Vallée's new categories in turn parallels the standard close encounter categories: sighting, physical effects, living entities, reality transformation, and lasting injury (numbered 1–5). Vallée also gives a credibility rating to each UFO experience.

Thus, Vallée's new classification system involves the following:

1. *Anomaly* (AN 1–5): The observance of a UFO and other accompanying or similar abnormal phenomena (for example, poltergeists).
2. *Flyby* (FB 1–5): The observance of a UFO flying in the sky.
3. *Maneuver* (MA 1–5): The observance of a UFO that exhibits a discontinuous trajectory in the sky.
4. *Close encounter* (CE 1–5): Same as Hynek's classification for close encounters.

With either Vallée's or Hynek's classification system, ufologists can begin to organize the broad diversity of UFO reports. This helps them sift through the reports to determine whether a natural explanation can be found for the experience or whether something really extraordinary has been going on.

## NATURAL EXPLANATIONS

MOST UFOS (unidentified flying objects) become IFOs (identified flying objects). In fact, John Spencer, British ufologist and the editor of *The UFO Encyclopedia*, asserts, "It is important to note that over 90 percent, possibly 95 percent, of UFO reports received are turned into IFOs and explained satisfactorily."[14] Most ufologists would agree with his assessment. Some would even put the percentage of IFOs at 99 percent or higher.

Ufologists spend much of their time determining the causes—apart from extraterrestrial or supernatural beings—of UFO sightings. They've come up with a number of standard explanations that adequately explain the preponderance of UFOs in terms of natural or normal occurrences.

The following sections offer eight natural explanations that can turn most UFOs into IFOs.

*Misidentified natural phenomena.* UFO reports often result from the misidentification of seemingly mysterious, but nevertheless natural, phenomena. The Center for UFO Studies (CUFOS) comments concerning misidentified natural phenomena:

> People report natural or conventional objects as UFOs because they do not recognize them as such due to unusual environmental conditions, ignorance, or the rarity of a natural event. For example, people have reported the planet Venus as a UFO, unaware of how bright the planet can appear at certain times of the year. Stars near the horizon are sometimes reported as UFOs because atmospheric turbulence and thermals (columns of warm air) cause them to twinkle rapidly in red and blue colors. Stars may also appear to dart back and forth because of autokinesis. This is a psychological phenomenon in which a person's eye movements create the illusion that a bright object seen in the dark without frame of reference is moving.[15]

Some of the natural objects or effects that commonly stimulate UFO reports include the following:[16]

- the moon, stars, and planets (for example, the cusps of a rising crescent moon in the tropics, Venus at maximum brightness, the Pleiades, and star clusters)
- unusual weather conditions (such as lenticular cloud formations, noctilucent clouds, rainbow effects, and high-altitude ice crystals)
- comets
- meteor swarms
- near or large meteors
- flocks of birds (sometimes carrying phosphorescent dust on their bellies and wings)
- swarms of flying insects
- reflections from atmospheric inversion layers
- hot ionized gas (natural or man-made)
- Earth lights (luminous electrical events from low-level

earthquakes and tectonic-geological phenomena)
- ball lightning
- reflected light (especially through broken clouds)
- aurora borealis (the northern lights)
- sun dogs (lights on either side of the sun caused by atmospheric distortions)
- red or green flashes (false solar images seen before sunrise or after sunset in high-altitude regions)

*Misidentified man-made phenomena.* UFO reports may be provoked by the following man-made stimuli:[17]

- balloons (meteorological and passenger)
- military aircraft
- unconventional aircraft or secret, advanced technology (for example, the SR-71 Blackbird or the B-2 Stealth bomber)
- aircraft with unusual external light patterns
- advertising planes
- artificial earth satellites
- hovering aircraft (such as a helicopter)
- blimps
- rockets and rocket launches
- kites
- fireworks
- lasers aimed at the clouds
- searchlights

*False images resulting from high-quality instruments.* Even the best instruments, including radar scopes, will produce ghost images under certain conditions. Every surface of a lens system can pick up reflections. Because many cameras have eight or more such surfaces, the potential for false image generation is significant. Any optical aids of the witness (eyeglasses, binoculars, and so on) must also be considered as a possible cause of a false image.

*False images resulting from faulty instruments or harsh environments.* Human error in the operation of instruments or mechanical failures that produce false images can result in UFO reports. False images may result from false radar readings (reflections, interference, indiscriminate detection) and photographic distortions (superimposed images caused by reflected light from such things as windows and eyeglasses). The environment sometimes causes false images. For example, photographic emulsions respond differently to environmental conditions. A cold, dark, drizzly day will produce a "UFO" in one emulsion, whereas a hot, dry, bright day will generate a "UFO" on a different emulsion. Also, if the camera's film is mechanically stressed or stretched, it can produce a variety of false images.

*False images resulting from faulty human perception.* Problems associated with human perception, especially optical defects (including distortions of sight, floaters in the eye, optical illusions, and mirages), may generate UFO reports. The most common false report in this category arises from the lack of adequate dark adaptation on the part of the observer. If fewer than ten minutes have passed between the witness's exposure to bright light and his or her nighttime UFO observation, a high probability exists that the UFO is a false image.

*Hoaxes, pranks, or fraud.* Deliberate deception can prompt UFO reports. The types of UFO hoaxes are many; examples include the following:[18]

- false or embellished testimony
- faked photography
- disk-shaped objects artificially suspended in the sky
- radio-controlled or programmed UFO look-alike craft
- cleverly crafted balloons, rockets, and models
- landing site frauds
- alien crash site and retrieval frauds
- fraudulent abductee and contactee claims

Concerning UFO hoaxes, CUFOS notes the following:

Although tens of thousands of UFOs have been reported over the past forty years, less than 1% have been shown to be hoaxes. For the most part, competent UFO investigators have been able to recognize hoaxes almost immediately. The most common type of UFO hoax is a prank balloon, which involves tying a flare or candle to a helium-filled balloon. On rare occasions elaborate hoaxes have been perpetrated, necessitating a more extensive investigation.[19]

Thus, while hoaxes, pranks, and frauds should be considered when a suspicious claim is made, this explanation is less common than most others.

*Nonempirical, subjective causes.* Sometimes UFO reports have no basis in objective, empirical reality but instead result from a broad range of inner and subjective factors, including the following:[20]

- neurological stimuli (for example, temporal lobe dysfunction)
- sociological stimuli (for example, cultural creations, myths, or conditioning)
- psychological stimuli (for example, psychosis, hysteria, sleep deprivation, paranoia, hallucination)

These factors are the province of social scientists. John A. Saliba explains how social scientists typically approach the subject of UFOs:

> Social and behavioral scientists have ... raised several issues regarding the meaning of UFO sightings and encounters, issues which have hardly surfaced in popular literature and are usually ignored or downplayed by ufologists. They have changed the more customary focus of UFO investigations—which is to verify UFO reports—by suggesting that since the UFO problem is not likely to be resolved in the near future, there is more to be gained by examining their sociopsychological significance. The meaning of the flying saucer phenomenon might lie more in its social and psychological dimensions than in whether extraterrestrials exist or not, or in what the aliens themselves are supposedly saying and doing. In other words, belief in flying saucers and alleged encounters with their occupants might reveal something important about human nature, the study of which is central to the social, psychological, and psychiatric disciplines.[21]

*Combinations of natural and human factors.* UFO reports may be initiated by a combination of natural factors. For example, a geophysical event may trigger a physiological response in the observer, which then may cause a psychological experience (hallucination). Canadian neuropsychologist Michael A. Persinger argues that many UFOs can be explained in this manner.[22] Another example is an aircraft that may appear unusual under uncommon weather conditions. Yet another example is the way the light of the sun may reflect off an aircraft's fuselage, creating a strange image.

## Two Leading Hypotheses

ALTHOUGH THE VAST percentage of UFO reports can be explained naturally, a residual percentage remains. These are called "RUFOs" for "residual UFOs." If only 1 percent of UFO reports remain unexplained, the number of RUFOs sighted over the last five decades could range into the tens of thousands, if not many more. How are these seemingly inexplicable UFO reports to be approached?

Some say that all UFO reports are simply the result of yet undiscovered natural factors, whether these are misidentified natural phenomena, misidentified man-made phenomena, sociopsychological factors, or a combination thereof. Cornell astronomer and popular scientist-writer Carl Sagan reflected this view when he said, "Although it is not possible to prove that all UFOs are misapprehended natural phenomena, there are no compelling reasons to believe otherwise."[23] Philip J. Klass, prominent skeptic and UFO debunker, argues that all UFO sightings and encounters are subject to natural explanations.[24] James Oberg, a scientist with NASA, asserts that it is not up to the skeptic to prove that every UFO case has a natural cause, but instead the burden of proof rests with those who advocate an extraordinary or supernormal explanation.[25] UFO skeptic Donald Menzel goes so far as to dismiss ufology as a "pathological science."[26] Swiss psychologist and psychiatrist Carl G. Jung argued that UFOs are not objective realities but instead represent a modern myth that arises from humankind's collective unconscious.[27]

Though many natural scientists hold to this explanatory hypothesis, credible and scientifically trained ufologists are among its critics. The following summarized objections deserve consideration: First, the presumption that all UFOs have natural explanations is a form of question begging that may prejudice any honest attempt to get at the facts. Second, the argument that extraordinary claims require extraordinary evidence may itself be an unproven assumption. Third, many UFO reports stubbornly resist a natural explanation even after being analyzed by competent and unbiased researchers. Fourth, viewing all UFO reports as misidentified natural phenomena may be the simplest hypothesis, but does that hypothesis best account for the facts and avoid unwarranted presumptions? Here, two glaring questions must be raised: (1) Has an objective, logical, and scientific analysis been made of the best-evidenced UFO cases? If not, why not? (2) What type of evidence (kind, quantity, quality) would be required to substantiate the reality of UFOs, and how would one justify such evidentiary demands?

Another view indicated by inexplicable UFO reports is that the natural, normal explanation of UFOs lacks conclusive explanatory power or scope, and thus there arises the necessity for a broader explanation for some UFOs. While numerous mystical explanations have been proposed for UFOs, ufologists identify two distinct "otherworldly" hypotheses: the extraterrestrial hypothesis (ETH) and the interdimensional hypothesis (IDH).

*The extraterrestrial hypothesis.* This view proposes that UFOs are objective, physical, and empirical realities—that is, metallic spacecrafts ("nuts and bolts") piloted by interplanetary space visitors.[28] These space entities, or aliens, represent a vastly advanced civilization (in technological and possibly moral and spiritual capabilities) that is presently studying humankind and will, at the appropriate time, make clear and unmistakable contact with humanity. Some advocates of the ETH also ardently believe that the United States government is covering up clear evidence of encounters with UFOs and extraterrestrial beings. In the words of one outspoken advocate of the ETH, Stanton Friedman, "The evidence is overwhelming that some UFOs are alien spacecraft."[29]

While the ETH remains the most popular explanation of UFOs,

especially among Americans, it is rejected as a credible hypothesis by many serious UFO researchers.[30] Key problems with the ETH will be addressed in the chapters to follow, but for now the following are ten summarized objections that critics of the ETH bring to bear:[31]

1. How are alien craft expected to traverse the vast distances of interstellar space, given the physical limits on how fast a craft can travel in space?
2. How can such a craft sustain a crew over these vast distances of space?
3. Why do sophisticated surveillance systems fail to detect incoming and outgoing UFOs?
4. How feasible is it for an extraterrestrial civilization, however advanced, to maintain a mission to Earth?
5. How is it possible that virtually every so-called metallic craft is different in size, shape, and color from the others?
6. Why are there so many different alien life-forms, and how do they readily adapt to space travel and to Earth's atmosphere and gravity?
7. Why do UFOs, as physical craft, not behave like physical objects but instead manipulate and violate the fundamental laws of physics at will?
8. What intelligent reason can be suggested for such bizarre and often absurd behavior as that exhibited by UFOs?
9. If alien visitors are physical, why do they so closely resemble or correspond to psychic or occultic phenomena?
10. As a proposed "advanced civilization" (in the areas of technology, morality, and spirituality), why do these aliens often comport themselves in a crude, sloppy, deceptive, and malevolent manner?

*The interdimensional hypothesis.* Like the extraterrestrial hypothesis, the interdimensional hypothesis purports that some UFOs are real phenomena that may exhibit physical and empirical effects.[32] But in this view the origin and nature of such phenomena belong not to extraterrestrial spacecraft but to another realm of reality beyond the time-space continuum. The IDH is thus sometimes described as the paranormal or occult view of UFOs.[33] Some ufologists (especially Christians) have ascribed an angelic or demonic interpretation to this interdimensional presence.[34] Even a number of leading secular ufologists have argued for a correspondence between UFO phenomena and the occult or demonism.[35]

People who hold this view differ as to exactly how one ought to define and explain this dimensional reality. Nevertheless, the IDH, while not a popular view in America, has support from some ufologists in countries around the world.

Critics of the IDH set forth the following summarized objections:[36]

1. What about the UFO reports that seem best interpreted as physical events, such as when UFOs appear on radar?
2. When some use the paranormal to explain UFOs, aren't they making an appeal for one unverified phenomenon to explain another unverified phenomenon?
3. What evidence is there for the existence of the paranormal that is superior to the evidence for the physicality of UFOs?
4. Doesn't the paranormal or occult hypothesis suffer from complexity, vagueness, and speculation?
5. Doesn't the demonic theory depend on accepting a narrow Christian perspective?

Determining which of these hypotheses promotes the strongest position poses difficulty, for all the reasons cited. That's why an examination of the extraterrestrial and interdimensional hypotheses from a scientific platform begins in the next chapter. The examination promises to provide fresh insight and lead the way toward greater understanding and a more satisfactory explanation.

CHAPTER 3

# LIFE ON OTHER PLANETS

*Hugh Ross*

It's five years in the future and the Extraterrestrial Life Search Foundation is in crisis. The foundation's leaders are finding it harder and harder to get funding. Even worse, their program has become a joke to many. After thirty years of operation, they have no success to show for their efforts.

Some of the foundation's leaders have gathered to discuss a plan they hope will bring success at last—or at least inspire enough hope so that they can get their credibility back. The plan is to conduct their search in a more targeted way by focusing on some of the hundreds of planets that astronomers have by then detected in orbit around distant stars. At a meeting they go through planetary profiles, looking for the planets that offer the best hope of supporting life. Someone brought in two boxes for the participants to put the profiles in when they are done discussing them; the boxes are labeled "Chance of Life" and "No Chance." At the end of the day the "No Chance" box is nearly full and the other one is as empty as at the beginning. As the meeting adjourns, the researchers slip out without meeting each others' eyes.

AUDIENCES MELTED when the lovable alien in Steven Spielberg's blockbuster film *E.T.* said in his sad, croaky voice, "E.T. phone home." But if extraterrestrials are really visiting Earth today, as many UFO enthusiasts believe, where is the home they would "phone" to? Are there lots of places in the universe where life could exist? Or maybe only a few? Or even only one—our own planet?

Films like *E.T.* and Carl Sagan's *Contact*, not to mention years of indoctrination in the worldview of naturalism, have primed the public to believe the fairy tale that life is spread all over the heavens. These words may seem like an indictment. They are not. Rather, they call upon scientists—and everyone else—to suspend for a time the assumptions of naturalism and to apply the methods of scientific inquiry. These methods can and do shed light on questions about UFOs and related phenomena.

Perhaps most importantly, scientific methodology builds boundaries, like fences, to corral the wild elements of the UFO mystery. Every reader already knows something about the scientific approach called "the process of elimination." Mechanics use it to find out why the car won't start. Doctors use it to find out why the stomach hurts. Detectives use it to find out who stole the cash. This process can also be used to discover what could, or could not, possibly give rise to UFO phenomena.

Ironically, some of the most obvious and potent challenges to the extraterrestrial hypothesis (ETH), as well as to extraterrestrial intelligence (ETI), have so far received little or no attention in UFO literature. This chapter, therefore, addresses one such challenge: Is there a place in the cosmos that a real and physical E.T. could call home? The answer to this question, via the process of elimination, requires that one carefully consider three issues: (1) the number of planets in the observable universe, (2) the probability that any given planet possesses essential life-support characteristics, and (3) the availability of sites other than planets where life might reside. This chapter will look at each of these issues. They provide a foundation for a rational evaluation of the ETH.

## The Number of Planets

AS RECENTLY AS ten years ago, astronomers were divided between those who believed that planets whirl around nearly every star in the cosmos and those who believed that the sun is a rarity in possessing planets. A couple of advances in science since then have, however, moved many astronomers from the former camp to the latter. First, while instruments and techniques have become available to detect planets orbiting other stars, not many have been found. Second, while sophisticated theoretical models have been developed to explain how disks of dust become planets, they don't support the conclusion that planets like those in this solar system are common. Both observationally and theoretically, therefore, it seems that solar-system-type planets are not so numerous after all.

Still, astronomers have discovered *some* extrasolar planets (planets outside Earth's solar system). Do these planets suggest that it's common to find planets that could sustain life? One must consider the evidence.

In 1992 astronomers first detected extrasolar planets. They found evidence of two small bodies (and later a third) orbiting the neutron star PSR 1257+12.[1] Because neutron stars are actually the crushed cores of supergiant stars that have exploded (that is, the remains of supernovae), the tiny bodies orbiting PSR 1257+12 cannot be called planets in the strictest sense. Rather than forming with the star, they must have formed as a result of, or sometime after, the star exploded. Or they might have been captured by the neutron star after the explosion. Their proximity to the neutron star means that they are constantly bathed in the deadliest radiation known to exist in our galaxy. Needless to say, they do not make a suitable home (or even neighborhood) for E.T.

In 1995 astronomers discovered a gas giant similar to Jupiter orbiting 51 Pegasi, an ordinary star located fifty light-years away.[2] Since then (and up to the time of this book's publication), astronomers have discovered sixty-six extrasolar planets orbiting other ordinary stars.[3] Of the sixty-six, one is a few times more massive than Earth, six are comparable to Saturn (Saturn = 95 Earth masses), and the rest are approximately the mass of Jupiter or greater (Jupiter = 318 Earth masses). Studying these planets gives astronomers a growing appreciation for the rarity of a planet with any substantial similarity to Earth, that is, any planet with life-support capabilities and gas giant partners similar to Earth's.

Each of the extrasolar planets discovered so far orbits a relatively young, metal-rich star (a star rich in elements heavier than hydrogen and helium).[4] This finding is no surprise. The heavy elements needed to make planets—and essential to life chemistry as well—require at least two generations of star birth, star death, and the scattering of star ashes. The principle of star formation is that the longer a galaxy sustains star formation, the more metal-rich its newborn stars will be. As for E.T.'s home, fewer than 2 percent of the stars in the Milky Way have access to enough heavy elements to produce planets that could possibly support life.[5] Roughly 98 percent are too old and too metal-poor.

Of the stars now known to have planets (stars within that 2 percent group), none are as old as the sun. Thus none can match the stability of Earth's sun and solar system, with its ensemble of rocky planets and gas giants in nearly circular, not too elliptical, orbits. The sun may be the only 5-billion-year-old star with adequate heavy elements to have formed both gas giants and small, rocky planets. The sun met with exceptional circumstances. It formed adjacent to a type I supernova and a type II supernova. Both blasted different arrays of heavy elements into the interstellar medium just previous to the condensation of the solar nebula (the cloud of dust and gas from which the sun formed).[6] Given the unusual convergence of all these critical factors for advanced life, and especially given the sun's age, the number of candidates for life sites within the Milky Way grows smaller still.

As for the availability of life sites in other galaxies, the odds look bleak. As it turns out, the Milky Way galaxy is exceptional for its number of late-born (young) stars. In some 94 percent of all galaxies, star formation shut down so long ago that they contain no metal-rich stars; hence, no planets.

The conclusion that only a small percentage of galaxies contain metal-rich stars and that only a few of those stars are old enough and metal-rich enough to possess planets has been confirmed recently by the Hubble Space Telescope (HST). A team of twenty-four astronomers focused the HST on a globular cluster of stars called 47 Tucanae. These stars are similar, physically and chemically, to the stars in most galaxies and to the vast majority of stars in the Milky

Way galaxy. (Globular clusters are the oldest pockets of star formation in any galaxy.) If planets were as common in 47 Tucanae as they are among the stars in the sun's galactic neighborhood, the HST survey should have detected seventeen—the number that mathematical modeling predicts. Instead, the research team found zero.[7]

Observations indicate that the number of stars in the cosmos with planets—any kind or size of planets—adds up to about a tenth of a percent of all stars. This number is at least a hundred times smaller than Carl Sagan's estimate that popularized the notion of extraterrestrial life and fueled the search for signals from intelligent extraterrestrial beings.[8] Yet, smaller though it may be, that number still adds up to a lot of planets. If each planet-possessing star has an average of ten planets, the number of planets in the observable universe would add up to $10^{20}$ (a hundred million trillion). But before one gets excited about the possibility of life on other planets, one must consider how likely it is that any of those planets offers a hospitable habitat for life.

## Hospitable Habitats

IN 1966 CARL Sagan argued that astronomers could use the statistics of this solar system to make grand-scale projections.[9] On that basis, Sagan estimated that approximately one planet in ten throughout the universe would manifest the conditions necessary for life. But given the advances in science since then, is his inference still valid?

One question hampered progress toward a realistic assessment of the number of possible homes for E.T.: To what degree does extraterrestrial life resemble life as we know it? At one time biologists speculated that extraterrestrial life might be based on exotic chemistry, not on carbon as earthly life is. Biochemists quickly determined, however, that the only elements other than carbon from which adequately complex molecules can be constructed are silicon and boron. Problem: silicon can hold together a string of no more than a hundred amino acids—far too short. Problem: everywhere in the universe, boron is less abundant than carbon, so carbon will always supersede it. Problem: boron in concentration is toxic to certain life-critical reactions. Physicist Robert Dicke long ago deduced that if one wants physicists (or any other life-forms), they must be carbon-based.[10]

The conclusion that all conceivable physical life-forms must be carbon-based permits scientists to develop an extensive list of planetary characteristics that must fall within a limited range for a planet to be capable of life support. And that list goes beyond the planet itself to the planet's star, moon or moons, planetary companions, and galaxy. This list grows longer with every year of new research. It started with two parameters in 1966, grew to eight by the end of the 1960s, to twenty-three by the end of the 1970s, to thirty by the end of the 1980s, and to forty-one in 1995.[11] The current list includes more than 140 parameters. (A sampling of the parameters that must fall within a certain narrow range for the support of physical life appears in appendix A.)

Each of these parameters must fit within certain limits to avoid disturbing a planet's capacity to support physical life. For some, including many of the stellar parameters, the limits have been determined quite precisely. For others, including many of the planetary parameters, the limits are less precisely known. After all, trillions of stars are available to study, while only sixty-six planets have been detected to date.

People must consider how confining these limits can be. Among the least confining would be the number of stars in the planetary system and the distribution of the planet's continents. The limits here are loose, eliminating perhaps only 20 percent of the relevant candidates. More confining would be parameters such as the planet's rotation period and its albedo (reflectivity), which eliminate about 90 percent of the relevant candidates. Among the most confining of all would be parameters such as the parent star's mass and the planet's distance from its parent star, where about 99.9 percent of all relevant candidates are eliminated.

Of course, not all the listed parameters are strictly independent of the others. Dependency factors reduce the degree of confinement considerably.

On the other hand, many of these parameters must be kept within specific limits for long periods of time. Given how variable the environments sometimes can be, this longevity requirement increases the degree of confinement. Life is indeed fragile.

The environmental requirements for life to exist depend quite strongly on the life-form in question. The conditions for primitive life to exist, for example, are not nearly so demanding as they are for advanced life. Also, it makes a big difference how active the life-form is and how long it remains in its environment. On this basis, there are six distinct zones or regions in which life can exist. In order of the broadest to the narrowest, they are as follows:

- Zone 1 — unicellular, low-metabolism life that persists for a brief time period
- Zone 2 — unicellular, low-metabolism life that persists for a long time period
- Zone 3 — unicellular, high-metabolism life that persists for a brief time period
- Zone 4 — unicellular, high-metabolism life that persists for a long time period
- Zone 5 — advanced life that survives for a brief time period
- Zone 6 — advanced life that survives for a long time period

And then there are some complicating factors to figure in. Unicellular, low-metabolism life is more easily subject to radiation damage and has a very low molecular repair rate. The origin-of-life problem (see chapter 4) is also much more difficult for low-metabolism life.

A calculation of the probability for there existing just one naturally occurring planet anywhere in the observable universe with the capacity to support physical life is outlined in appendix B. That probability is less than 1 chance in $10^{174}$ (the number 1 followed by 174 zeros). To put that number in perspective, the entire universe contains only $10^{79}$ protons and neutrons, and every reader has a much higher probability of being killed in the next second by a failure in the second law of thermodynamics (about one chance in $10^{80}$).

These statistics have driven some scientists to abandon the premise that E.T. requires an Earthlike home. They have begun to speculate about other locations, called "exotic life sites," where life might exist in the universe.

## EXOTIC LIFE SITES

COULD LIFE EXIST on a satellite (moon) of a planet instead of on the planet itself? Could life exist on a planet that is traveling freely through space instead of orbiting a star? These are a couple of the questions that scientists today are asking.

*Moon home.* Some scientists speculate that a satellite orbiting a giant planet (which in turn orbits a star resembling the sun) at a distance nearly identical to Earth's distance from the sun could be a life site.[12] But this alternative can be tested against a long list of recent findings, and it does not fare well. Here are just a few of the problems:

- None of the sixty-six planets found to date outside Earth's solar system orbit their stars within the zone required for life support.[13] This finding comes as no surprise, because large planets form under cold, low-radiation conditions far from stars. By gravitational interactions with the interplanetary media (dust, comets, asteroids, and other planets), most drift inward toward their stars. Such planets that do not drift in too close to their stars, however, lack the stable, nearly circular orbit which life demands.[14] Of those sixty-six planets, the few that orbit nearest to the life-habitable zone have such highly eccentric—that is, elongated elliptical—orbits as to make life on their satellites (if they have satellites) impossible.[15] The question remains unanswered as to whether or not giant planets can retain their satellites as they migrate inward.

- A satellite orbiting close enough to its planet to avoid enormous seasonal temperature fluctuations becomes tidally locked—the same side always faces the planet. To avoid day-night temperature extremes, the satellite must orbit very close to its planet. However, the tidal forces from such a close orbit would cause a host of life-destructive effects, from huge climatic instabilities, to massive and frequent volcanic eruptions (such as those on Jupiter's moon Io), to orbital instabilities.[16] The tides from a close orbit would exert torques so great as to cause the satellite to move farther and farther away from its planet.[17] Any possible life-favorable conditions would last briefly, at best.

- A satellite with a life-sustaining atmosphere (highly improbable in the first place) would likely lose it in short order. Charged particles emanating from the planet's magnetosphere (the area in which the planet's magnetic field traps and dominates the behavior of charged particles) would blast it away—unless that satellite somehow possessed a strong magnetic field (similar to that of the sun, Jupiter, and Earth). Jupiter's moon Ganymede—the largest known planetary satellite and the only one with undisputed magnetism—has a magnetic field less than 1 percent the strength of the Earth's.[18] Given such a weak field, neutral water vapor clouds (essential to a life-sustaining atmosphere) would never form there.[19]

- Another risk for a satellite closely orbiting a large planet is that the planet's gravity would significantly attract asteroids, comets, and other debris passing near it. This enhanced flux would mean increased bombardment, and such bombardment would be catastrophic to any possible life on the satellite.

- The satellite cannot retain an adequate atmosphere for life unless its mass exceeds 12 percent of Earth's mass.[20] At the same time, the satellite needs a mechanism to compensate for the star's increasing luminosity (brightness, thus heat radiation) as it ages.[21] The only known mechanism is the one seen on Earth, called the carbonate-silicate cycle.[22] This cycle will not operate, however, without lots of dry land (which eliminates ice-water environments such as that of Jupiter's satellite, Europa), without cryptogamic colonies (soil-conditioning colonies of bacteria and fungi),[23] without vascular plants,[24] and without plate tectonic activity similar to Earth's (which requires at least 23 percent of Earth's mass).

*Starless planet.* Sustaining the quest for other potential life sites, planetary scientist David Stevenson and origin-of-life researchers Jeffrey Bada and Christopher Wills have speculated that life might not require a home near a star. They suggest this scenario: A planet may be ejected from a normal planetary system before losing any of its light gases. If so, the planet may have enough surface warmth (from interior radioactive decay) and a sufficiently heavy atmosphere of molecular hydrogen to sustain life chemistry and metabolism.[25]

Does this hypothesis fare any better than the satellite hypothesis? In fact, this hypothetical life site (made less probable by the rarity of radioactive elements) might serve as a brief stopover for primitive extraterrestrial life. But it would not last long enough within the life-support range of temperature and other conditions to serve as a home for intelligent E.T. And devoid of the protection provided by a nearby star and gas giants, a far-flung planet would suffer life-exterminating bombardment by comets.

This exotic-life-site hypothesis, like the other, appears to be an interesting yet ultimately fruitless suggestion for how intelligent life might exist elsewhere in the universe. The bottom line is that, if E.T. has a home, it must be a planet like Earth orbiting a star like the sun in a galaxy like the Milky Way. And that possibility, as ongoing research shows, seems less possible as each year passes. In fact, the number of known characteristics that must be fine-tuned for physical life has more than tripled since 1995. Meanwhile, the probability for finding a planet, or other heavenly body, anywhere in the universe with the capacity to support life has shrunk by a much greater proportion. E.T. does indeed appear to be homeless—unless, of course, a transcendent, supernatural Being built a home for him.

But the lack of a home for E.T. is just part of the problem facing those who argue that UFO phenomena can be explained by the activities of physical extraterrestrial beings. As the next chapter suggests, the origin of life may well present a far greater challenge.

CHAPTER 4

# Evolution's Probabilities

*Hugh Ross*

At about the same time that planet Earth was forming, an important event occurred on another planet in a nearby galaxy. That planet was then much like the Earth was later to become—small but stocked with a diversity of elements and having a hardened crust covered in part by water. One day, in a tidal pool near one of the planet's seas, a lightning strike set off a subtle yet fateful conversion: some of the complex molecules in the warm pool of water recombined in a way that changed the material from nonliving to living.

You might not have believed it if you had looked into the tidal pool that day, but there, cupped in the rock, lay the whole future history of life on the planet—simple organisms, fish, amphibians, birds, and eventually advanced, intelligent beings, all just waiting for the universal forces of evolution to do their work. And while the Earth went through its own development and evolutionary process, the intelligent beings on the other planet would develop their society and their technology and become curious about the universe surrounding them. In time their curiosity would lead them to build craft that could carry them to other life-bearing planets, including ours.

A likely scenario?

ACCORDING TO deeply entrenched naturalistic beliefs, the existence of life on Earth implies that the right mix of simple hydrocarbon molecules in the right pond or mineral-rich mud, sparked by lightning or some other energy input, can and will—over the course of time—spontaneously produce one or more simple organisms. Given still more time, these simple organisms will evolve along natural pathways into advanced plants and animals. Since this process occurred on Earth, the theory goes, it can occur elsewhere in the cosmos on some of the billions and billions of planets out there.

The previous chapter established how unlikely it is that any planet other than Earth could support life. Now one must consider whether, even if such a planet existed, life could develop there by purely natural means. If the answer is yes, then the extraterrestrial hypothesis (ETH) for explaining UFOs might be true. If the answer is no, then of course the ETH is a dead end.

The simplest way to evaluate the plausibility of a naturalistic origin-of-life scenario for E.T. is to evaluate the plausibility of that scenario where life exists already: on Earth. That's just what this chapter does.

Some years ago Richard Dawkins wrote a popular book called *Climbing Mount Improbable*.[1] In it he claimed that a series of inevitable events led to the seemingly improbable conclusion of life's appearing on Earth naturally. But the reality is that decades of research findings converge to show how truly improbable, indeed impossible, are the "inevitable" events Dawkins describes.

In this chapter these four scientifically established conclusions are examined: (1) The Earth has not existed long enough for life to have arisen by natural means. (2) The conditions were not right on early Earth for life to begin naturally. (3) Life could not have come to Earth from space. (4) Life is too complex to have arisen by natural means on Earth or anywhere else as described in evolutionary theory.

It's not looking good for the extraterrestrial hypothesis.

## A Matter of Time

THE ENORMOUS COMPLEXITY of even the most basic life-form defies description. Even more so does the leap from inorganic to organic, from nonliving to living. Given how slowly existing life-forms change, naturalists have acknowledged that life's origin requires time—eons of time. This is true even if one accepts paleontologist Stephen Jay Gould's hypothesis that changes in life-forms occur much more rapidly during periods of extreme environmental stress. And that is why a 1992 paper sent shock waves through the science community, the ripples from which continue to this day.

In 1992 Christopher Chyba and Carl Sagan published a review paper on the origins (yes, plural) of life.[2] Using their own and others' research, they argued that life began, died out, and began again repeatedly on Earth before it finally took hold. This suggests a much more rapid appearance of life than would seem possible, given the processes that evolutionary theory describes.

This rapidity is seen, for example, in the way life appeared very early in Earth's history. Lunar meteorites confirm that the Earth's crust remained molten until 3.9 billion years ago.[3] Yet fully formed cells show up in the fossil record as far back as 3.5 billion years ago, and limestone (composed of organic remains) dates back at least 3.8 billion years. In other words, forms of life existed within 100 million years of the Earth's surface cooling. But life goes back even further in time than that. The ratio of certain carbon molecules—specifically, carbon-12 to carbon-13—found in ancient rocks indicates that life abounded on Earth as early as 3.86 billion years ago.[4] Thus, primitive life appeared on Earth within a time window no wider than 40 million years.

While 40 million years might seem a long time to most people, naturalists consider it hopelessly minuscule. And new findings suggest that the time window was even narrower. The era between 3.86 and 3.5 billion years ago held grave dangers for life, including bombardment by asteroids (some huge enough to be called small planets) and comets.[5] The intense bombardment that prevailed from 4.3 to 3.9 billion years ago gradually subsided between 3.9 and 3.5 billion years ago, but astronomers calculate that at least thirty catastrophic impacts must have occurred during the latter period. In other words, life sprang up on Earth (and then sprang up again) in what could be called geologic "instants"—periods of 10 million years or less—between impacts.[6]

Some optimistic researchers initially suggested that the bombardment could have contributed to life's origins. Glossing over the destructive effects of the collisions, they proposed that this bombardment might have delivered concentrated doses of biochemical building blocks from extraterrestrial sources. Further research argues against this idea. Though some comets, meteorites, and interplanetary dust particles do carry simple hydrocarbons (molecules containing two of life's key elements) and even a few amino acids, they carry far too few to make a difference. Nor are the amino acids all oriented left-handed as life requires. A slight excess of left-handed over right-handed amino acids in a few meteorites has been proved to have arisen from terrestrial biological contamination.[7] In fact, with every helpful molecule would have come several more unhelpful ones—useless molecules that would have gotten in the way of the needed ones.

The conclusion one must draw is that bombardment from space destroyed the fragile web of life on Earth repeatedly—and that life repeatedly reappeared within a time frame much too short to make room for evolutionary processes. And yet, even if there had been more time, would Earth's conditions have been right to allow the evolution of life?

## THE SOUP'S NOT ON

MIDDLE SCHOOL AND high school science textbooks published as recently as 2000 still lean on the worn-out hypothesis that life came together in a primordial "soup" (warm ponds and wet mineral surfaces enriched with life-building molecules).[8] Even under the highly engineered, highly favorable conditions of a laboratory, however, such soups have failed to produce anything remotely resembling life. At best they produce only a random distribution of the very simplest of life's building blocks.

Life chemistry demands that its nucleotide sugars be "right-handed" (having their hydrogen molecule on one and the same side) and that most of its active amino acids (nineteen of twenty) be "left-handed" (having their hydrogen molecule on the opposite side). Despite decades of research and quantum leaps in technology, lab researchers cannot come close to lining up molecules with the correct handedness. Neither can they assemble them in the correct sequence to make life. The futility of expecting a bunch of simple molecules to bring themselves together into a functioning, living organism in just a few million years in the chaotic conditions of early Earth by far exceeds that of expecting a spilled bowl of alphabet soup to spontaneously generate a poem.

The ratio of certain carbon molecules (specifically, carbon-12 and carbon-13 isotopes) found in ancient rocks reveals more than just the early appearance of life on Earth. That ratio also helps researchers distinguish between inorganic carbon-containing molecules (those that form life's critical building blocks, or prebiotic molecules) and similar molecules that result from the decay of once-living organisms (postbiotic molecules). Detailed analysis reveals that even the most ancient carbon-containing molecules are all postbiotic; none are prebiotic. In fact, neither a primordial, prebiotic soup nor a mineral-rich, moist layer ever existed on Earth.

The recently discovered "oxygen-ultraviolet paradox" helps explain why no such soup or substrate existed: The existence of oxygen in the atmosphere and the ocean would guarantee the shutdown of prebiotic chemistry. The absence of oxygen, on the other hand, would allow intense ultraviolet radiation to penetrate Earth's atmosphere and upper ocean layer, also guaranteeing the shutdown of prebiotic chemistry. Either way, the primordial soup explanation for the origins of life utterly fails.

Despite the abundance of life on Earth today, this planet begins to look more and more like a place that was never suitable as a place for life to develop by naturalistic means.

## LIFE FROM SPACE

BECAUSE EARTH CONDITIONS defy any naturalistic origin of life, some researchers have turned their hopes to the skies. Either from elsewhere in this solar system or from more distant precincts of the universe, they suggest, life may have come to this planet. From that point on, evolution could have done its thing on Earth, jumpstarted by the gift from outer space.

*Life from the Red Planet.* In recent years a number of news reports surfaced about the possibility of life having once existed on the home of fiction's original little green men: Mars. Martian life in a simple form might have been transported to Earth, some scientists propose, when asteroid and comet impacts on Mars sent Martian rocks hurtling through interplanetary space, with life attached, to land on Earth.

Conjectures about life from Mars have been loudly and widely heralded in the media. Images taken by unmanned Mars explorers show that Mars was once, briefly, both warm and wet, though it never had a precipitation-fed water cycle.[9] With this news, the familiar mantra of naturalists comes into play: "Where there's water, there's life." However, the same factors that rule out Earth as a fortuitous cradle for life also rule out Mars. These factors include the destructive effects of intense bombardment and the oxygen ultraviolet paradox, described above.

Theoretical support does exist, however, for the opposite of the Mars-to-Earth life scenario.[10] The same processes that bring Martian debris to Earth also deliver Earth debris to Mars. About 2 percent of all the material ejected from Earth's surface into space because of comet or meteor collision inevitably falls on Mars. Over the past nearly 4 billion years, several million kilograms of Earth-life remnants must have been deposited on Mars. It seems only a matter of time before researchers detect—and possibly misconstrue—its presence.

*Life aboard space rubble.* With Mars holding no realistic promise as a source of life, theorists next considered comets and meteorites. Not claiming that life arose on or in comets and meteorites, some researchers have suggested that these kinds of space rubble could serve as taxicabs, carrying prebiotic molecules picked up in interstellar space safely to Earth.

There is, however, an insuperable problem with this suggestion. Because conditions on early Earth would not have permitted any natural pathway for simple prebiotic molecules to assemble into organisms, the meteorites and comets would have had to deliver advanced prebiotics, such as proteins and DNA and RNA molecules. But no trace of such molecules, nor of the complement of amino acids, nucleotides, and sugars necessary for life, has been found in recovered meteorites. Furthermore, using the chemical classes of compounds found in the Murchison meteorite (the largest recovered meteorite), origin-of-life researcher Robert Shapiro showed that side reactions would effectively prevent any simple prebiotic molecules that might be present from ever spontaneously forming into more complex molecules, such as proteins, PNA (peptide nucleic acid), DNA, or RNA.[11]

*Life in particles borne on stellar "wind."* With the odds stacked so heavily against any naturalistic origin-of-life scenario in the environs of this solar system, some researchers are dusting off Fred Hoyle's panspermia proposition from the 1970s.[12] This hypothesis says that life originated somewhere else in the cosmos and came to this solar system on star-generated "wind."

All stars manifest radiation pressure—light intense enough to push tiny particles through interstellar space. But light that intense would include enough ultraviolet radiation to kill a microbe in a matter of just a few days. That's why many scientists have considered Hoyle's proposition a failure.

Some scientists have suggested that if a microbe were encased inside small dust grains, it might be protected from the radiation. But this proposal also fails.[13] To move a more massive dust grain requires more intense starlight. The ultraviolet and x-ray radiation in such starlight would penetrate the dust grain and kill the microbe inside. Furthermore, the only source of light strong enough to move a dust grain encasing a microbe is a supergiant star. For a variety of reasons, from radiation effects to gravitational effects and more, life cannot possibly arise in the vicinity of, or survive anywhere near, a supergiant star.[14]

At the 2001 Lunar and Planetary Science Conference, Dr. Jay Melosh from the Lunar and Planetary Laboratory and the University of Arizona reported on the feasibility of Earth's capturing an interstellar wandering rock (that is, a piece of an asteroid, comet, or planet).[15] His studies demonstrate that, at best, Earth could capture only about $3 \times 10^{16}$ of the available material per year. To put it another way, Earth has only one chance in ten thousand of capturing just one interstellar sample at any time in its entire history. Thus the transport of significant pieces of interstellar dirt, let alone life, to Earth is virtually impossible.

The impossibility of transport represents just one of the intractable problems facing the panspermia hypothesis. Another has to do with where "out there" in the cosmos life could naturally arise and survive. As chapter 3 notes, astronomers can point to no site in the universe suitable for life sustenance, much less origination. They cannot, for example, find a place where all the amino acids are left-handed and all the nucleotide sugars are right-handed.[16]

## THE COMPLEXITY PROBLEM

WHEREVER LIFE ARISES, by far the biggest problem is the unfeasibility of generating, without supernatural input, the required degree of complexity. A wide gulf separates an aqueous solution of a few amino acids from the simplest living cell.

Years ago, molecular biophysicist Harold Morowitz calculated the size of this gulf. If one were to take the simplest living cell and break every chemical bond within it, he said, the odds that the cell would reassemble under ideal natural conditions would be one chance in a number so big that to write it out would take thousands of pages.[17]

In light of such a number, the time scale issue becomes completely irrelevant. What does it matter if every possible planet in the universe has been around for 10 seconds, 10 billion years, or 10 quadrillion years? Even if all the matter in the visible universe were converted into the building blocks of life, and even if, by some unknown means, assembly of these building blocks proceeded randomly once every microsecond for the entire age of the universe, the odds would improve by a barely perceptible fraction.

One might think that naturalists would be hopelessly discouraged by Morowitz's odds. But that is not the case. They have searched for ways to bring the odds more in favor of their belief that life arose naturally on Earth and probably on other planets as well.

*Response 1: sequencing flexibility.* Naturalists typically counter Morowitz's odds by pointing out that not every amino acid and nucleotide must be strictly sequenced for life molecules to function. They are right about that, and their observation does improve the odds a bit.

However, the odds go the other way when one learns that Morowitz treated *all* the amino acids as bioactive (participants in life chemistry). In fact, only twenty of the more than eighty naturally occurring amino acids are bioactive, and of that twenty, only the left-handed ones enter in. Furthermore, Morowitz assumed totally favorable conditions and totally constructive chemical processes. Under natural circumstances, destructive chemical processes operate at least as frequently as constructive ones. Thus the odds for the assembly of the simplest living entity actually grow worse as more details are figured into the calculation.

*Response 2: simplicity.* Some scientists suggest that the simplest living entity 3.5 billion years ago may have been far simpler than the simplest organism of today. If they are right, that would change Morowitz's calculations.

The difficulty here is that conditions on Earth 3.5 billion years ago were insufficiently different from conditions today to support such a hypothesis. In astronomical terms, conditions are similar enough that one would expect spontaneous generation of life to continue today if it occurred then, and evidence of that is not seen.

But the simplicity (or rather, reduced complexity) of early organisms presents another problem. Organisms below a certain level of complexity cannot survive independently. Complete genome sequences of the oldest and simplest independent life-forms appearing in the fossil record—life-forms 3.5 billion years old—contain between fourteen hundred and nineteen hundred gene products (genes describing the assembly sequence of functional proteins).[18] It seems early organisms were not, and could not have been, as simple as some scientists have suggested.

*Response 3: multitalented RNA.* A few papers recently published in the journal *Science* discuss what seemed, at first glance, a possible way around some of the complexities of life's origin.[19] Provided is a brief review of the background to explain the researchers' hopes— and disappointments.

Molecules responsible for life chemistry cannot function by

themselves. DNA (molecules that hold the blueprints for constructing life molecules), proteins (molecules that follow portions of the blueprints in building and repairing life molecules), and RNA (molecules that carry the blueprints from the DNA to specific proteins) are all interdependent. Thus, for life to originate mechanistically, all three kinds of molecules must emerge spontaneously and simultaneously from inorganic compounds. Even the most optimistic of researchers agree that the chance appearance of these incredibly complex molecules of exactly the right type and number at exactly the same time and place lies beyond the realm of natural possibility.

A ray of hope that this complexity barrier could be overcome came from an experiment performed in 1987. It demonstrated that one kind of RNA can act as an enzyme or catalyst (an agent to facilitate a chemical process). That is, this RNA molecule appeared to perform the functions of a protein, at least to a limited degree.[20]

This finding led to some leaps of faith. Because researchers already assumed that RNA can come together under prebiotic conditions more easily than DNA or proteins can, some wondered whether a primitive RNA molecule, capable of functioning as a protein *and* as DNA, evolved by natural means out of a primordial soup. In time, it was proposed, this "primitive" RNA specialized, evolving into the three kinds of molecules now recognized as RNA, DNA, and proteins.

Newer discoveries showed still more protein-like capabilities of RNA molecules. A research group presented evidence that a certain RNA molecule could stimulate two amino acids to join together with a peptide bond (the kind of chemical bond formed in proteins).[21] A second research team observed another RNA molecule both making and breaking the bonds that join amino acids to RNA.[22] Though these capabilities, plus the ones observed earlier, add up to only a tiny fraction of all the functions proteins perform, several origin-of-life theorists proposed that no proteins were necessary for the first life-forms.

These new findings and hypotheses may seem to streamline the origin of life by natural processes, but a closer look at the realities say they do not. Even if a single primordial molecule could perform all the functions of modern DNA, RNA, and proteins, such a molecule would be no less complex in its information content (that is, its built-in "knowledge" of what to do) than the sum of modern DNA, RNA, and proteins. In other words, the task of assembling such an incredibly versatile molecule is no easier than assembling the three different kinds of molecules. The information content of the three is simply concentrated into one enormously complex molecule. Even Leslie Orgel, a leading proponent of an RNA origin of life, admits, "You have to get an awful lot of things right and nothing wrong."[23]

Another catch in these arguments is the false notion that RNA assembles more easily than do proteins or DNA. For twenty years researchers and texts taught that RNA had been synthesized in a lab under prebiotic conditions. Robert Shapiro exposed this myth at a meeting of the International Society for the Study of the Origin of Life held at Berkeley, California, in 1986. Shapiro traced all references to RNA synthesis back to one ambiguous paper published in 1967. He went on to demonstrate that the synthesis of RNA under prebiotic conditions is essentially impossible. Shapiro later published his case against RNA self-synthesis in the journal *Origins of Life and Evolution of the Biosphere,* and it is a case that remains unchallenged to this day.[24]

Additional problems put the self-assembly of both Earth life and extraterrestrial life in doubt. One is RNA survivability. The various naturalistic hypotheses for life's origin demand that RNA and its components hold together for many millions of years. At warm temperatures RNA sequences come apart like beads on a broken string. At the time of life's origin, Earth's surface was hot—roughly between 80° and 90° C (176° and 194° F), without any cold spots.[25] The RNA nucleotides themselves (the beads) decompose at warm temperatures in just a few years or even, according to some results, in a few weeks.[26] Even at cool temperatures, problems exist. Cytosine (one of the RNA nucleotides) decomposes in less than seventeen thousand years if kept as cool as 0° C (32° F).[27]

Outside the protection afforded by a cell membrane (composed of two layers of complex fatty molecules with a bunch of specialized proteins sandwiched in between), no environment provides sufficient stability and protection for RNA molecules and their nucleotide bases. In other words, RNA molecules cannot survive outside cell membranes, while cell membranes cannot be constructed without RNA. Both must come together simultaneously.

With such observations, the RNA response to Morowitz's odds appears completely inadequate. The life span of the universe, much less the life span of any particular planet in the universe, is far too short to have permitted life to arise on its own anywhere at all.

## Time to Give Up?

IN THE FACE of increasing, rather than decreasing, challenges to their viewpoint, the origin-of-life community of scientists has begun to lean hard toward directed panspermia—the hypothesis that some unknown intelligence within the cosmos sent life to Earth by some unknown means, for some unknown purpose, to colonize it. But the research potential here looks bleak and unpromising.

Physicist and author John Horgan laments the situation in these gloomy terms: "Science in its grandest sense—the attempt to comprehend the universe and our place in it—has entered an era of diminishing returns.... Scientists will continue making incremental advances, but they will never achieve their most ambitious goals, such as understanding the origin of the universe, of life, and of human consciousness."[28]

Such gloom can be dispelled, however, by a willingness to question the presuppositions of naturalism. The scientific method demands a new hypothesis when the old one fails to produce significant results. What Horgan observes is not the end of science; rather, it is the limitation of a particular model for the origin of the universe, of life, and of human consciousness—a model that rules out, dogmatically, any possible connection between nature and supernature.

Life, after all, does exist on Earth; so, where did it come from? It must have come from an intelligence not within but beyond the universe. That, at any rate, is the view of the authors of this book. However, this book is not the place to consider in any depth the subject of the divine creation of the universe and of life on Earth.

The subject of this book is lights in the sky and little green men, and there's still one more nail to hammer into the coffin of the extraterrestrial hypothesis. Even if planets elsewhere in the cosmos could support life, and even if life could arise on those planets naturally in the first place, it would still be physically impossible for intelligent life from other planets to travel here in spacecraft for purposes scholarly or warlike.

CHAPTER 5

# INTERSTELLAR SPACE TRAVEL

*Hugh Ross*

At the spaceport of the Intergalactic Research Facility on the planet Qaxdljnik, Org and his team are ready to board their spacecraft. They've been assigned to study the alien inhabitants of a far-distant world—a rocky yet water-rich planet, the third in its solar system. Org waves the three fingers of his right hand at his wife and children for the last time, then walks on his spindly legs into the craft, a circular ship barely twenty of Org's paces in diameter. Now all he has to do is travel 230 million light-years, avoid deadly space radiation and debris, and survive for tens of thousands of years—long past the life expectancy of his species on Qaxdljnik—just to reach the alien world. But to observe these "humans" on this "Earth" going about their daily routines, it's worth it, right?

AS VIEWERS OF the television show *Star Trek* watched the *Enterprise* zoom through space at warp speed week by week in the 1960s, they pondered the possibilities of space travel. Then astronauts landed on the moon, and space travel seemed still more realistic. Humans proved space travel is possible. And if humans can travel through space, why not aliens?

What if, by some miracle of chance, a planet with all the features for life support does exist somewhere in the cosmos, maybe even nearby? What if, by an even greater miracle, life did arise on that planet and develop a curiosity to explore space, to find Earth, and to communicate with Earthlings? What kind of technology would these extraordinary extraterrestrials need to successfully make the trip? The more visual reader might ask, "What would alien spacecraft look like?"

This chapter addresses what it would take to travel across the vast reaches of space. But first, a common misconception needs correction. NASA's routine successes in sending spacecraft to Earth's sibling planets can cause people to lose sight of the hard limits governing interstellar or intergalactic space travel by intelligent physical beings. Such limits have little to do with technological capability. They have much more to do with the laws and constants of physics, which no amount of technological capability can overcome.

## THE SCALE OF SPACE

TO COMPREHEND THE vast distances between stars seems beyond the capacity of human imagination. To say that the nearest star is 25 trillion miles away simply does not make the point. An analogy, or scale model, may help, at least a little: if the sun (nearly a million miles in diameter) were represented by a grapefruit, and if someone were to place that grapefruit in the middle of downtown Los Angeles, then a second grapefruit representing the nearest star would lie in downtown Managua, Nicaragua.

Given such distances, the question "Are we there yet?" takes on new meaning. If E.T. were to catch a ride on the fastest spacecraft ever built by NASA, that craft would take 112,000 years to reach Earth's solar system from the nearest star. What if the alien craft were much faster than anything NASA has produced? Even traveling at half the velocity of light, a spaceship would take nearly nine years to make the trip.

Realistically speaking, the trip would have to be much longer than the trip from the nearest star or stars. Sentient physical beings require an Earthlike habitat, one that orbits a single, middle-aged star closely resembling the sun. This planet's orbit must be nearly circular, not too eccentric. The planet must be shielded from asteroid and comet bombardment by a massive companion planet but cannot be bounced around by the gravity of that protector planet. Many more criteria could be listed,[1] but these make the point. Astronomers can say with the assurance of long and careful observation that no star within about fifty light-years of Earth meets the requirement of having a gas giant nearby.[2] Those stars with mass similar to the sun's are either too young or too old to burn with sufficient stability for life support. Some possess partner stars or huge nearby planets that would disrupt the orbit of any Earthlike planet. The rest are lacking a gas giant planet that could shield an Earthlike planet from too many asteroid and comet collisions.

Even if intelligent beings were to reside a mere fifty light-years distant, they would have to cut a zigzag course through various galactic hazards to reach planet Earth, making their trip considerably longer. Incoming travelers would have to dodge the gravity and deadly radiation of neutron stars, supergiant stars, nova and supernova eruptions, and the remnants of such eruptions. They would have to avoid the gas, dust, and comets that are so dense in the spiral arms of this galaxy, as well as the environs of late-born stars (stars formed during the past 5 billion years). But they would have to stay in the plane of the galaxy. Any departure from the plane would expose the travelers to the deadly radiation that streams from the galactic core. Maneuvering to avoid hazards would extend the minimum distance to an estimated seventy-five light-years.

Recent findings, however, push that minimum figure even higher. Based on the assumption that any interplanetary craft would likely maintain communication with the home planet or with other members of the traveling party, a SETI (search for ETI) research group scanned all 202 of the roughly solar-type stars within 155 light-years of Earth. Not one intelligible signal was detected anywhere within the vicinity of these stars.[3] This finding translates to a minimum alien travel distance of 155 light-years plus hazard-avoidance maneuvers—a total of roughly 230 light-years (or 1.35 quadrillion miles).

One significance of these enormous distances lies in the fact that distance translates into time, and time translates into risk exposure. The more time a living or mechanical body spends in space, the more dangers it encounters—deadly dangers.

## The Speed of Travel

AS SOME READERS may remember from their high school science classes, the laws of physics forbid any chunk of matter from traveling faster than the velocity of light. Serious difficulties arise, however, long before an object reaches that speed. At the velocity of light, the energy required to move a specified mass is infinite. At even half the velocity of light, the energy needed to propel an object is 170 million times greater than NASA's fastest spacecraft requires.

The energy problem compounds, however, because propellants and engines themselves involve mass. The higher a spacecraft's speed, the more propellant and the bigger the engines it requires. Therefore, the higher the intended speed of the spacecraft, the (exponentially) higher the mass of the craft. An additional mass problem arises, of course, from the need to move the spacecraft's payload (the total weight of the passengers, crew, instruments, and life support supplies). The mass of a craft and its propulsion system rises geometrically relative to the mass of the payload.

The need for speed poses yet another problem: the faster an object travels through space, the greater the damage it suffers from collisions with space debris. In fact, the increase is geometric. Micrometeorites (fractions of a millimeter across), for example, punched holes the size of silver dollars in the Hubble Space Telescope's solar panels. If the telescope had been moving a thousand times faster (relative to the micrometeorites), the damage would have been a million times worse.

In terms of space debris, though, micrometeorites may be the least of a space traveler's worries. Computer modeling indicates that there is a large cloud of comets, estimated to contain 100 billion comets or more, surrounding the solar system. Such clouds likely surround any star in this galaxy that could possibly harbor planets. Astronomers suspect that the giant molecular clouds scattered throughout the Milky Way galaxy may contain even greater numbers of comets.

To protect against damage from space debris, a spacecraft needs some kind of armor. However, armor means more mass, which means more propellant to move the added mass. More propellant means more propellant to move the extra propellant. Thus the problem escalates.

While space debris poses a lesser risk at lower velocities, lower velocities also mean greater travel time. The probability of damage from space debris rises in proportion to the amount of time spent in space—and it rises with the *square* of the velocity. Therefore, in terms of damage from debris, space travelers face deadly dangers at any velocity, slow or fast. And slow or fast, a spacecraft will suffer general wear and tear to its component parts.

Exposure to radiation poses yet another serious threat. The faster a craft travels through space, the greater the damage it suffers from radiation. The particles associated with radiation (for example, protons, neutrons, electrons, heavy nuclei, and even photons) cause erosion to the "skin" and components of the craft. Again, the rate of erosion rises with the square of the velocity. However, a slower velocity means more time in space, and that extra time means more radiation exposure for the aliens on board. (No matter how thick any safety shield may be, some radiation inevitably leaks through.)

By conservative estimates, any reasonably sized spacecraft transporting intelligent physical beings can travel at velocities no greater than about 1 percent of the velocity of light. At higher velocities the risks from radiation, space debris, leaks, and wear and tear simply become too great to prevent the extinction of the space travelers before they reach their destination. A spacecraft traveling at 1 percent of the velocity of light (nearly 7 million miles per hour) would need twenty-three thousand years to travel 230 light-years.

## WORMHOLES

HIGHLY IMAGINATIVE AND technically trained space travel enthusiasts suggest that advanced aliens may have found a way to use space-time "wormholes" to travel to distant locations in the universe in a relatively short time. On closer examination, however, this idea offers no help in solving the distance and time problems.

General relativity says that massive objects distort the curvature of space and time in their vicinity. The greater the mass-density of an object, the greater the degree of space-time curvature it produces in its immediate vicinity. General relativity predicts that when matter becomes sufficiently compressed by its own gravity (as in a black hole), a discrete region of space-time will develop where the curvature becomes infinitely sharp. That is, a "singularity" will develop at the center of the mass concentration.

If a black hole connected to one sheet of space-time in the universe happens to make contact with another black hole connected to a different sheet of space-time, that point of contact may (hypothetically) offer a travel corridor. The point of contact, however, must be singularity to singularity (see figure 5.1, diagram 2) so that a traveler

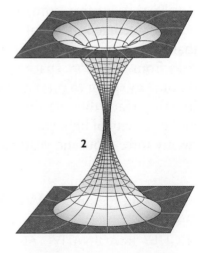

Diagrams courtesy of Reasons to Believe

Fig. 5.1. These diagrams represent black holes, singularities, and wormholes. (1) Massive objects, such as black holes, bend the curvature of space and time near them. At the core of a black hole, a region of infinite mass density and space curvature (a singularity) would develop. Any alien or spacecraft approaching a black hole would be stretched into a string of particles extending for miles. (2) Black holes could, in theory, connect at their singularities and form a space-time "wormhole." Some nonphysicists have speculated, therefore, that space travelers might be able to pass from the space outside another black hole. However, not only would the travelers be destroyed as they approached a black hole, but not even the smallest physical particle could pass through the wormhole. (3) Because wormholes are (in theory) made from bent space and time, the entrances to the black holes creating the wormhole would typically be relatively close together. Therefore, not much space and time would be saved in taking the wormhole shortcut.

funneling into the center of one black hole can come into contact with the center of another black hole.

While these so-called wormholes connecting one black hole to another black hole are mathematically possible, one must question the physical practicality of their use by alien travelers. According to the best-established models for the universe, regions of space that could be connected via wormholes are already close to one another. In other words, the use of a wormhole would offer little time advantage. One cosmic model in which a ten-dimensional space-time sheet bends to make a U (see figure 5.1, diagram 3) offers the possibility of a significant shortcut through space, but ongoing research has yet to verify the viability of such a model.

From a realistic perspective, the wormhole question is moot. Because black holes are relatively rare, the probability that the singularities of two spatially distant black holes would come into contact is virtually zero, as is the probability that any possible wormhole would lie in a location useful for alien space travel. General relativity dictates, further, that wormholes would be extremely unstable. The longevity of any possible wormhole is far too brief to allow any physical being to travel through.

The most devastating evidence against wormhole travel comes from the application of physical laws to discrete physical objects. In short, any physical object would be destroyed beyond recognition by the intense gravitational forces in the vicinity of a wormhole. An unlucky alien approaching one would be stretched into a long line of particles several miles long. As this alien gets sucked toward one of the singularities, even the particles would disintegrate. The alien would end up as an extremely compressed ball of chaotic energy.

While wormholes are mathematically possible, they offer no hope or help for alien travel. Not even the tiniest physical entity could survive passage through a wormhole. Coming anywhere close to a wormhole means destruction.

## Multigenerational Travel

AT TWENTY-THREE thousand years (minimum) for a one-way trip

from home to Earth, space aliens would no doubt face some daunting social challenges. Life spans anywhere within the confines of the universe must be finite, not infinite, according to the laws of physics. Moreover, life spans inevitably decrease with exposure to radiation such as space travel yields. The complexities of carbon-based biochemistry (the only possible chemistry for physical life) set life's limit at about a thousand years, even if traveling aliens were to hibernate for long periods.

A journey across more than 230 light-years would extend through multiple generations. A multigenerational journey presents another set of difficulties. Whether or not the original voyagers volunteered for the mission, their progeny would receive the mission by inheritance, not by choice. Like it or not, the mission is theirs. If space travelers were to resemble humans in any way, one can easily imagine that dedication to the original goals might be difficult to maintain from generation to generation. Changed or confused priorities would likely add to the trip's duration, among other difficulties. They might even lead to a turning back.

Multigenerational survival for space travelers requires a sufficiently large and diverse base population of initial passengers. Otherwise, the aliens would most likely become extinct before their craft reaches its intended destination. And a population of any size, from two to twenty thousand, requires various resources and systems for sustenance. At a minimum, these resources and systems must include some sort of food, respiration products, and waste recycling, all maintained at sufficient levels to minimize the risk of ecological disaster.

A one-way trip that takes twenty-three thousand years or more raises additional doubts about the alien travelers' survivability. The extinction risks, given the limited population and all the contingencies of space travel, seem overwhelming. As humanity has discovered during the past fifty years, a civilization advanced enough to launch a trip through space may not survive long enough even to build a transport and get it off the ground. High technology carries a terrible price: reduced survivability.

High technology and resultant high living standards mean individuals carrying deleterious mutations typically survive long enough to reproduce. High technology and high living standards strongly encourage both men and women to delay reproduction. In a high-tech world, an individual needs more time to be educated and trained for self-sufficiency—and even more for making a contribution to ongoing technological advance.

Delayed reproduction, particularly for males, results in transmission of increased numbers of deleterious mutations.[4] According to one research study, the human population at the close of the twentieth century suffered an accumulation of deleterious mutations measuring three per person per generation.[5] This rate significantly accelerates humanity's movement toward extinction.

To make matters worse, wealth and technology inversely correlate with the birthrate. In other words, the greater a society's wealth and the greater its use of technology, the fewer offspring it produces. Today, not a single nation with a per capita income exceeding $20,000 has a birthrate high enough to prevent eventual extinction. In Europe and Japan, for example, the birthrate is less than 75 percent of that needed to maintain the population at a constant level.[6]

For space travelers, all these problems are compounded by limits on the size of their traveling party. Whereas 6 billion people living on a large planet can tolerate epidemics, natural disasters, ecological crises, and wars, a few (or few thousand) individuals on board a spaceship or cluster of spaceships would likely be wiped out by such catastrophes. Humanity holds the added advantage of a large habitat with a wide variety of refuges where one can find temporary escape from a given problem or disaster.

These extinction risks suggest that for distant stars and planets, technology sufficient for space travel is much more likely to doom a society's destiny than to fulfill it. Intelligence would tell such aliens to stay at home or to limit their colonization efforts to their own planetary system.

## Machines Versus Aliens

OBVIOUSLY, THE CONSEQUENCES of damage by space debris, radiation, leaks, ecological breakdown, and wear and tear are much worse for intelligent physical beings on board spacecraft than for mechanical instruments. If gathering (or giving) knowledge is the goal, living beings typically have the advantage over machines in that they can adapt more quickly and successfully to changing circumstances and unexpected contingencies. However, as the travel distance increases, the advantage shifts: the difficulty of transporting live travelers grows and the travelers become less adaptable relative to machines.

Even for exploration of Earth's solar system, machines hold a huge advantage. For visiting the moons of Jupiter (less than 0.0001 light-years away), at least ten thousand instrument missions can be sent for the cost of one manned mission. If something goes wrong with an instrument on such a mission, no one dies (though someone may lose a job). If the instruments detect something they were not designed to probe, another set of instruments can be designed and sent out. If circumstances warrant a longer stay, it can be accomplished with a little redesign or extra provisioning, in most cases. A few intelligent beings on a month-long mission are likely to learn much less about a planet or a moon than would ten thousand space instruments operating over many years.

This kind of analysis would not be lost on aliens more advanced than humans. If aliens exist on distant planetary systems and have some interest in planet Earth, they would more likely send machines than members of their own species.

This conclusion, like the other ones regarding the physics of interstellar space travel, is discouraging to the extraterrestrial hypothesis (ETH). Not only is it statistically unimaginable that another planet capable of supporting life exists in the universe and that life developed there naturally, but even if there were such a place, intelligent aliens living there could not reasonably be expected to cover the vast distances from their home to ours. At this point it's best to scratch the ETH.

But where does that leave the curious in explaining UFOs? Are all UFO sightings just misperceptions? It might be nice if it were that simple, but it is not. There is a small but irreducible number of UFO sightings that cannot be explained away. While alien beings from outer space are not behind UFO phenomena, there may be something real going on here after all.

CHAPTER 6

# RUFOs—THE UNEXPLAINED UFOs

*Hugh Ross*

At a scientific lecture in Kiev, a burly Russian scientist rose to his feet in the middle of the lecture hall, complaining first in Russian, then in English, "Oh, no! Another scientist here to tell us UFOs aren't real!" Five hundred of his peers held their breath and turned toward the podium to see how the speaker, an American astrophysicist, would respond.

"You misunderstand me," he replied. "I can only demonstrate that UFOs defy the laws of physics. That is not to say they cannot be real."

"Explain yourself, and I will listen," said the Russian, who promptly sat down.

MANY INTELLIGENT well-educated people have difficulty dismissing all UFO phenomena as illusory. Ordinary, sane people—not just the proverbial "crackpots"—encounter UFO phenomena, and they know they have seen or experienced something out of the ordinary. Among those who have not personally witnessed a UFO, some listen to the UFO reports of others and, using their normal powers of assessment, become persuaded of the credibility of some of them.

Respected ufologists agree that there must be something real at the bottom of some UFO reports. The work of these people has generated a new acronym: "RUFOs" for "residual UFOs." The term refers to those UFOs left over after all others have been explained away. These residuals lie at the heart of virtually all UFO questions and controversies. These sustain the mystery.

## The Prevalence of RUFOs

UFO RESEARCHERS AGREE that most so-called UFO phenomena may be more accurately dubbed IFOs (identified, or identifiable, flying objects). While statistics on the total database are not available, analysis has been done on official UFO reports.

Computer scientist and astronautical researcher Jacques Vallée made a careful study of more than twelve thousand UFO reports (including the Project Blue Book files) submitted to the U.S. Air Force between the years 1948 and 1969. This compilation of files represents a screened list; obvious hoaxes and misidentifications have been filtered out. Vallée's careful scrutiny ruled out more than three-fourths of the reported phenomena as IFOs—astronomical bodies, airplanes, balloons, rockets, satellites, and so on. However, 23 percent (nearly three thousand) could not be identified as having any possible natural or human-devised source.[1]

Astronomer J. Allen Hynek, founder and director of the Center for UFO Studies (CUFOS) and an official consultant for Project Blue Book, roughly agrees with Vallée's figure, estimating that about one in five UFO reports lack any recognizable natural or human explanation.[2] Alan Baker, author of the *Encyclopedia of Alien Encounters*, estimates, more conservatively, one in ten are unidentifiable.[3] Other astronomers with significant personal experience in responding to and studying UFO reports suggest lower percentages, but none propose 0 percent.

The difference in estimates may be attributable, at least to some degree, to psychosocial factors. While a person may be eager for the attention of an official UFO investigator, he or she may be understandably reluctant to face the skepticism of an astronomer. Astronomers, in fact, often receive reports secondhand, from the friends and family of reluctant witnesses. People also differ in their feelings about potential media exposure. Some welcome it; others (arguably the majority) fear it. Both Vallée and Hynek speculate, on the basis of their in-depth research, that UFOs lacking in any identifiable physical origin must be more numerous than statistics reveal.[4]

However many RUFOs there may be, enough have been investigated to begin drawing some conclusions about what they are like.

## Reality Checks

MORE THAN A dozen research scientists have reviewed the RUFO database, taking care to eliminate hoaxes, frauds, and individual hallucinatory experiences.[5] Many physical effects remain without physical explanation.

Perhaps the most objectively researched consequence of RUFOs is physical trauma. Soil, plants, animals, humans, and machines in the vicinity of reported UFO landing and crash sites have suffered disturbance or damage. In addition to soil depressions consistent with described "landing pods," researchers have cataloged some eight hundred scorched, denuded circles of land related to landing marks or depressions.[6] According to one investigator, the general pattern is either "a circular patch, uniformly depressed, burned, or dehydrated," or a narrow ring, from one to three feet wide and with a thirty-foot (or larger) diameter.[7] Some patches and rings have remained barren for weeks, months, or even years, and tracks as deep as five to eight inches have been confirmed.[8] The soil compression, including crushed rock, at one reported landing site matches that of a thirty-ton object.[9] Heat indicators range from melted snow to brittle or molten rock, calcined materials, metallic slag of unusual composition, and even altered soil chemistry and rock chemistry.[10]

Plants and trees at some sightings have appeared scorched or blighted.[11] In some cases, these plants and trees took more than a year to recover from the trauma. In one unusual case, grass and weeds at the site grew to double or triple the size of identical plants in adjacent areas and maintained this exceptional growth for two years following the sighting.[12] In another case, nearby witnesses confirmed that a hovering UFO, which seemed to emit neither heat nor sound, cooked and desiccated the plants beneath it.[13]

At Trans-en-Provence, France, on July 8, 1981, UFO investigators found more than damaged leaves. They found that the biochemistry of UFO-marred leaves possessed markers of extreme aging, and so did the very young leaves that sprouted after the sighting.[14] Michel Bounias, the research toxicologist who performed the study, ruled out chemical poisoning as a possible cause. The one viable possibility—powerful microwave radiation—could produce some, but not all, of the biochemical consequences. The results do not resemble anything known to exist, natural or man-made.

Animals, especially dogs and cattle, but also rabbits, goats, and horses, have shown noticeable agitation in the presence of UFO phenomena.[15] These animals have reportedly reacted before their human observers did, not simply in response to the humans' body language. In some cases, incessant barking and mooing occurred before, during, and after the sighting. In other cases, dogs reportedly cowered and refused to go outside at the time of a UFO event, and cattle herds stampeded.

Apart from the abduction and sexual assault claims common to close encounters of the fourth kind, the physical symptoms most frequently reported by human observers of RUFOs include nausea, headache, temporary paralysis or blindness, numbness, and sensations of heat or weightlessness at the time of the event.[16] Disturbed dreams have followed some observers for weeks, and family members and friends confirm radical changes in the observers' beliefs and values.

Some observers have experienced serious injuries and even death.[17] Many have required hospitalization for burns and wounds. Such injuries have, on occasion, been accompanied by hair loss, diminished vision, diarrhea, internal bleeding, liver damage, and swelling. Vallée documents twelve cases of fatal injuries, most victims surviving less than twenty-four hours after the UFO encounter.[18]

Machines and electrical systems have also been impacted by RUFOs.[19] Most commonly, vehicle engines, radios, and lights have been observed to turn off during a UFO sighting and then return to normal operation afterward. In a few cases, car batteries overheated and quickly deteriorated. Diesel engines have rarely been affected. Airline pilots encountering UFOs have noted disruptions in radar, radio, and compass operations.[20] In one case, that of a United Airlines DC-10, three cockpit compasses all gave different readings—a circumstance that should have alerted the mismatch annunciator system to shut down autopilot. It did not. Upon landing, all three compasses returned to normal.

In one well-documented UFO event, a helicopter climbed upward though the pilot held the controls fully in the downward position.[21] Witnesses both in the helicopter and on the ground confirmed this phenomenon. They also reported that as soon as the UFO disappeared the pilot regained full and normal control.

A peculiar physical effect associated with RUFOs is called "angel hair."[22] Certain low-flying, rapidly zigzagging UFOs have been observed to leave extensive trails behind them. According to witnesses, the trail material slowly fell to the ground, sticking to trees, bushes, telephone wires, and roofs, much like spider webs. The white strands have been described as long (up to sixty feet) or short (mere specks). They have been photographed but never retained for investigation. Though reportedly picked up by numerous witnesses, the angel hair disappeared quickly after being handled. Untouched angel hair also disappeared quickly; some observers described it as wafting up into the sky and vanishing.

Given all these observed, measured, and corroborated physical effects, it might seem as though RUFOs have been confirmed as physical reality. But in fact, that is not so.

## Unreality Checks

IN FILMS or TV, spacecraft are free to fly without any limitations except those of their creators' imagination. But in the real world, the laws of physics apply. It seems evident that RUFOs must be nonphysical because they disobey firmly established physical laws.

Unlike physical entities, RUFOs typically exhibit the following characteristics:

1. RUFOs leave no physical artifacts, even after crashing.
2. They generate no sonic booms when they break the sound barrier, nor do they show any evidence of meeting with air resistance.
3. They may be seen but not photographed, or they may be photographed (though never with high resolution) but not seen. In fact, the resolution of a UFO image may change from one moment to the next.
4. RUFOs may be detected by radar but not seen, or they may be seen but not detected by radar.

5. They make impossibly sharp turns and sudden stops and impossibly rapid accelerations to speeds approaching fifteen thousand miles per hour.
6. RUFOs hover aboveground or harm buildings and trees without any movement of air—no downward rush or other movement counter to ambient air currents.
7. They change momentum without yielding an opposite change of momentum in matter or in an energy field either coupled to the object or in the vicinity of the object.
8. They change shape, size, and color at random.
9. RUFOs suddenly disappear and reappear, or they disintegrate and reintegrate.
10. They send no detectable electromagnetic signals.
11. They emit light that casts no shadows. They project light beams of finite length or emit some light that twinkles and other light that does not. They change the apparent color of people, objects, or vehicles they spotlight.
12. They sometimes remain indistinguishable in shape despite close observation.
13. RUFOs consistently succeed in evasive action, sometimes vanishing instantly or at other times seeming to enter the ground without leaving a trace.
14. They melt asphalt and metal objects, and burn grass and leaves, without fire or flame.
15. They physically injure and even kill human observers apart from any identifiable physical agent.

RUFOs' improbable predictability, the similarity of their physical manifestations, attests to their reality on one hand, but that same quality of predictability also suggests their nonphysical nature. For example, relative to the number of people out of doors and capable of making observations, ten times as many sightings occur at 3:00 A.M. as at either 6:00 A.M. or at 8:00 P.M.[23] Many more RUFOs appear in remote areas than in built-up or populated regions. They seem to prefer lonely roads to fields or villages. The witness list for each RUFO is small, rarely more than six people.

The likelihood of a person's witnessing a RUFO bears no correlation to that individual's number of hours spent watching the sky. People trained in night sky observation see no more RUFOs than others. If RUFOs were indeed physical, then pilots, air traffic controllers, astronomers, and other sky watchers, especially those with the most hours logged, should see more RUFOs than those with less time spent looking up, but that is not the case.

Based on a preponderance of evidence that RUFOs cannot be physical, the National Aeronautics and Space Administration (NASA) turned down President Jimmy Carter's request to undertake UFO research.[24] In its 1977 statement to the president and the nation, NASA leaders affirmed the agency's research mandate, one that focuses on the natural realm, exploring physical phenomena subject to the physical laws and dimensions of the universe.[25] Due to the absence of evidence supporting RUFOs as physical entities, they said, NASA could not engage in UFO research.

This NASA announcement inspired a wide range of reactions. Unable to merely walk away from the abundant, if strange, evidence that RUFOs have some kind of reality, many people accepted the notion that the government wants to hide something. Those with a penchant for conspiratorial thinking went even further in their speculations. These responses demonstrate how powerfully the human mind battles the seeming contradiction between something's being demonstrably real and its also being demonstrably nonphysical. In later chapters a resolution of that contradiction will be offered. But first, a look will be taken at whether there is anything to the claims of a government cover-up of UFOs.

CHAPTER 7

# GOVERNMENT COVER-UPS

*Mark Clark*

After successfully passing through exhaustive security protocols, Lt. Col. John McPeel has finally been given clearance to join the staff at the Air Force's unacknowledged Groom Lake facility in Nevada. As he rides in a jeep toward the site for his first day of work there, he wonders what he'll find at the infamous Area 51. Is it really just a development site for experimental military aircraft, as he's been told? Or are the rumors true? Is this where the government houses flying saucers recovered from crash sites in New Mexico? Will he join a decades-long program focused upon producing weapons by reverse engineering from alien technology?

The final gate has just opened and he will soon find out for himself.

TRUE BELIEVERS in UFOs are frustrated at not being able to convince the world that a residual portion of UFO sightings are for real. Could someone be making their task more difficult? The only entity that would seem to have both the scientific capability to acquire knowledge about UFOs and the power to contain that knowledge is the U.S. government. And so the loaded word "cover-up" enters the conversation.

Mention government cover-ups and three incidents invariably come to mind for those who study UFO phenomena. The first is the supposed crash of an alien spacecraft near Roswell, New Mexico, in 1947. The second is the Air Force's Project Blue Book (also known

80   Lights in the Sky and Little Green Men

as the Condon Report), which investigated and rejected the reality of UFOs. The third is Area 51 (also called Groom Lake), the site in Nevada where the government purportedly holds evidence of UFOs.

UFO conspiracy theorists insist that all three offer proof of U.S. government deception. And yet all of them may have simpler, more accurate explanations than conspiracy theorists would have people believe.[1]

These three famous examples will be considered as test cases for whether there could be anything to the claims of government cover-up.

# Roswell

THE MANY STORIES and myths that developed around the 1947 Roswell incident make separating fact from fiction difficult. UFO researchers compounded the difficulty by changing the story.[2] Recently, after analyzing the major works on Roswell, a team of scholars from Brandeis University detected six variations of the incident.[3] Beginning in 1980 and continuing through 1996, major parts of the legend have been changed or altered to accommodate new information, discredit some of the wilder claims, and promote forgeries (for example, the Majestic-12 documents) to prove government involvement.[4] Over the years, new researchers have entered the field. UFO researchers themselves differ widely on the credibility of some of these claims and often sharply disagree about their respective interpretations.

In mid-June 1947 something landed on the J. B. Foster sheep ranch operated by W. W. "Mac" Brazel, approximately seventy-five miles northwest of Roswell. (One account places the crash site 175 miles northwest on the plains of Saint Augustine.) Brazel first saw the material on June 14, ten days before the first modern report of flying saucers. On June 24, pilot Kenneth Arnold observed and reported nine disklike objects while he was flying near Mount Rainier in Washington. That report opened the floodgates to other sightings.[5] Brazel did not return to collect the material at the ranch until July 4. He then reported his findings to the local sheriff on July 7. Though he wouldn't have heard of flying disks on June 14, he certainly had become aware of the reports by the time he turned in the debris.

The physical evidence recovered from the Foster ranch included "aluminum-like" material, Scotch tape, some other tape with markings described as flowers or "hieroglyphics," wood, and other material weighing no more than five pounds. Photos of the recovered materials are widely available.[6]

Conspiracy theorists differ on whether a UFO landed, crashed, or bounced off the area, leaving some debris or even alien bodies behind. They also disagree on how many craft landed. In the early 1990s, after U.S. Air Force researchers discovered credible archival evidence for high-altitude balloon research, conspiracy theorists argued that the craft crashed into a balloon. The current story is that the disabled experiment produced the debris picked up at Roswell and that the "real" stuff was hidden by the government. One variation of the story claims that an autopsy was performed on an alien in the base hospital at Roswell Army Air Field (RAAF). Each researcher's variation of the story multiplied assumptions to explain away anomalous facts and the others' research.

A hasty press release issued by the base press relations officer inadvertently planted the seeds of the conspiracy story. Former bombardier 1st Lt. Walter Haut, apparently under orders from the base intelligence officer, Maj. Jesse Marcel, announced in the headline of a local newspaper that "RAAF Captures Flying Saucer on Ranch in Roswell Region."[7] Maj. Marcel took the material found near Roswell to the Eighth Air Force Headquarters in Fort Worth, Texas.

The next day, however, an official press release contradicted the flying saucer story. Warrant Officer Irving Newton, base weather forecaster, and Brig. Gen. Roger Ramey, Eighth Air Force commander, issued a press release declaring that the discovered items were remnants of a weather balloon. Two contradictory statements by military officials put the appearance of a government cover-up in place.

During this time, at nearby Alamogordo Army Air Force base, researchers from New York University (under contract to Watson Laboratories) were conducting a series of high-altitude balloon tests. All of the balloon flights were unclassified weather balloons. But part of the payload on some balloons was a highly classified research experiment designed to detect Soviet nuclear detonations acoustically.[8] Even the name of the project—Mogul—was a secret to those who worked on the project.

A senior scientist working on Project Mogul, Professor Charles Moore, said in a recent publication that he wasn't told either the name or the purpose of the experiment until 1992. However, he and others guessed its objectives as they worked on the project.[9] Reconstructing events from his notes and records, Moore identified the recovered Roswell debris from NYU Balloon Flight 4 found in June 1947. He noted that the balloons were equipped with corner reflectors for radar tracking and pieced together with balsa wood, Scotch tape, and tape from a toy company in New York. This tape had purple and green flower designs on it—"hieroglyphics," as later called by some.[10]

No one questioned the official explanation for over thirty years. Further, no UFO research tally of thousands of suspected UFO incidents in subsequent years ever mentioned Roswell.[11] More than thirty years later, UFO conspiracy theorists excavated the Roswell story by relying on shaky memories, hypnosis of some key "witnesses," and second-and third-hand testimony.

How do conflicting stories escalate into conspiracy theories such as the one about Roswell? A brief review of the government's need for secrecy and how it classifies information can help people understand how conspiracy theories get started.

## CLASSIFIED INFORMATION

IT'S IMPORTANT TO remember that in 1947 the Cold War was just beginning. Government officials were doing their best to learn the secrets of the Soviet Union and keep the Soviets from learning America's secrets. But in fact, the security of information vital to national interests was a concern before the Cold War and it remains a concern today, following the Cold War.

The U.S. protects classified government information with at least two methods. First, the information is assigned a level of secrecy from confidential to secret to top secret and beyond. At each step, more effective measures are taken to protect the secret and more rigorous checks are performed on those given access to it. Second, and more importantly, at higher levels of classification, information is divided into separate areas, called "compartmentalized" information. Depending on the sensitivity of the material, such intelligence is further compartmentalized under code words and may require a certain level of clearance for access.

Having a clearance, however, is insufficient reason for gaining access to classified information. A person acquiring a clearance also has to have a "need to know"—an official reason for such a clearance. Then, as the job requires, he or she may be "read" into a particular compartment. And when read into a compartment, he or she is also briefed on the responsibilities of having such access, including the legal penalties for violating that trust.

Scholars typically disdain security clearances—and for good reason. These clearances block the free flow of information among researchers that is crucial to successful scholarship. Moreover, security clearances may at times cause problems because the left hand doesn't always know what the right hand is doing. Officials say contradictory things because they have access to different information. To the public, this can make the officials appear to be lying. On the other hand, secrecy precludes the enemy from knowing what the government is doing. Thus, gaining a comprehensive look into U.S. intelligence capabilities or military secrets becomes more difficult for enemy spies. This added measure of security makes the challenges associated with secrecy worth the possibility of one or two key people's access to information being compromised.

The reason for confusion at Roswell is obvious. Both those in official capacities at the time and researchers since have come up against "compartmentalized" (highly classified) information. If Professor Moore, project director for Mogul, didn't know the name of his own project, then it is likely that few others did either, especially those not associated with the New York University balloon experiments. Certainly 1st Lt. Haut and Maj. Marcel at RAAF in Roswell, and probably Gen. Ramey and the weather forecaster at Eighth Air Force headquarters in Fort Worth, had no knowledge of the Soviet nuclear detonation detection experiments.[12] Their responsibilities lay elsewhere.

Marcel and Haut speculated that the recovered debris was the remnants of a crashed saucer. But rancher Brazel's enthusiasm and the general interest in the flying saucer phenomena bolstered the public's excitement. The official rebuttal by Gen. Ramey was that the debris was a weather balloon. He may or may not have concocted the weather balloon story as a cover to maintain secrecy. In any case, he was not far off the mark. But many UFO conspiracy theorists can take the simple truth and give it a completely new spin.[13]

## Project Blue Book

THE GENESIS OF the Air Force's Project Blue Book came swiftly after the 1947 Roswell incident, but not directly because of it. Project Blue Book evolved out of projects Sign and Grudge, which were smaller Air Force projects conducted in 1948 and 1949 to look into the rash of UFO sightings, which were beginning to demand explanations.

In the wake of the first significant wave of flying object sightings, which began in Washington with the pilot Kenneth Arnold in 1947, the Air Force launched its first official UFO research team. The Technical Intelligence Division of the Air Material Command at Wright Field (later Wright-Patterson) in Dayton, Ohio, began Project Sign (initially Project Saucer) in January 1948 to monitor all information concerning such sightings. The project's initial goals were modest. Although generally inquisitive about UFO phenomena, the group's principal concern was the possibility of the Soviet Union's developing secret weapons. Soon enough, however, Air Force researchers concluded that the vast majority of UFO sightings were real, explainable, and nothing extraordinary.

Shortly after concluding Project Sign, the Air Force launched Project Grudge. As Air Force researchers became convinced that UFO reports were not extraordinary phenomena, the new mandate sought to alleviate public anxiety and keep tabs on the reports of UFOs. More importantly, Project Grudge officials found no evidence of advanced or exotic Soviet weapons behind the UFO reports and concluded that UFOs did not threaten national security.[14] The project officials sought to explain the sightings by referring to natural events where appropriate: balloons, conventional aircraft, meteors, optical illusions, and similar natural explanations. Finally, Air Force officials terminated the project in December 1949, believing that its involvement might be contributing to the public scare and to a "war hysteria" about the Soviet Union.

Toward the end of the Korean War, a large number of reported sightings again stimulated official U.S. interest in UFOs. Reports of sightings around Washington, D.C., and nearby Andrews Air Force Base especially disconcerted the Truman administration. The U.S. Air Force director of intelligence, Maj. Gen. Charles P. Cabell, ordered a new study on UFOs, entitled Project Blue Book. The principal focus of the new Air Force investigation was to again alleviate public anxiety by showing that UFOs were naturally occurring phenomena, neither a threat from the Soviet Union nor a threat from outer space.

From the beginning, the Central Intelligence Agency (CIA) monitored the Air Force's work on UFOs. Possible threats to U.S. national security initiated its concern. But early in its involvement, the agency wanted to conceal its role from the media and the public. The agency accepted the Air Force's rationale that over 90 percent of reported sightings could be explained by naturally occurring phenomena, and it didn't want to inadvertently confirm the "reality" of UFOs.[15] The director of Central Intelligence, Walter Bedell Smith, assigned the agency to help the Air Force maintain objectivity in its investigation and to determine what role UFO phenomena might have in psychological warfare.[16]

In late 1952 the CIA established the Robertson panel to examine UFO phenomena on a national scale. A physicist from the California Institute of Technology, H. P. Robertson chaired the distinguished group of scientists. The panel reviewed Air Force records and reports to determine possible dangers to U.S. national security. It also found neither any evidence of a direct threat to the United States nor any evidence that the objects sighted were of extraterrestrial origin.[17]

The panel expressed concern that reports about UFOs would generate unjustified public hysteria and might hinder U.S. officials from protecting U.S. national security. The panelists feared public overreaction, because the evidence suggested that the objects were neither physical threats in themselves nor evidence of new Soviet weapons. Excessive reporting might clog communication channels and give the Soviets opportunities to exploit a manufactured hysteria and launch a surprise nuclear attack. The panel recommended that all U.S. governmental agencies downplay the mystery behind UFO reports and that an education program be developed for the public to help people distinguish real threats from imaginary ones.[18]

By 1955 a new problem arose for both the CIA and the Air Force. In August of that year the agency began its U-2 overhead reconnaissance flights (high-altitude surveillance missions). At the time, commercial aircraft flew no higher than twenty thousand feet and frequently less, but the U-2 reached sixty thousand feet. And because the first U-2s often reflected sunlight from that altitude after it was already dark on the ground, a new wave of UFO sightings occurred. CIA officials determined that over half of all UFO reports in the late 1950s through the 1960s were attributable to manned reconnaissance flights, especially over U.S. territory. Blue Book research officials, aware of the U-2 program and later the SR-71 Blackbird program, were obliged to explain the sightings by referring to naturally occurring phenomena, such as ice crystals and temperature inversions, so as to protect the secrecy of the highly classified reconnaissance flights.[19]

By the mid-1960s, the Air Force sought a way out of Blue Book.[20] More than fifteen years had elapsed since the project's inception in 1948, and the Air Force no longer needed to justify the importance of its existence to protect national security. A costly project, Blue Book drained Air Force resources. No evidence ever emerged that UFOs posed a threat to U.S. national security or that they were of extraterrestrial origin.[21] After several failed attempts to get universities to take over the project, the Air Force finally contracted with the University of Colorado for an eighteen-month study. A former director of the National Bureau of Standards, physicist Edward U. Condon, headed the project.

Condon's leadership of the university project has been heavily criticized by some UFO advocates as well as by a small faction of his own researchers.[22] J. Allen Hynek, a Northwestern University astronomy professor and scientific adviser to Blue Book before the Colorado University project, criticized the research methodology of the Condon team. Though a skeptic of Blue Book, Hynek argued that no government conspiracy occurred.[23] Later, he founded the Center for UFO Studies, a private group of researchers.[24]

In April 1969, Condon and his committee released their report, which concluded that further extensive study of UFO sightings was unwarranted.[25] After a special panel of the National Academy of Sciences reviewed and concurred with the Condon Report, the secretary of the Air Force, Robert C. Seamans, terminated Project Blue Book.[26]

## Bureaucratic Politics

THE U.S. AIR Force's research into UFO reports is a classic study in bureaucratic politics.[27] Political scientists developed the bureaucratic politics model to explain how, and often why, decisions in the government are made—and why they often appear irrational to outsiders.

All bureaucracies compete for scarce resources (that is, money). They tend to seek the national interest, but each sees a different face of an issue. The Air Force seeks to defend the U.S. by using planes; the Navy, by using ships. Because each service competes with the others for resources to accomplish national military goals, the Air Force was quick to get into the UFO research business. As a relatively new agency in 1948, the Air Force needed to establish a preeminent reason for its existence. The Soviet threat and air power's ability to deliver atomic munitions provided it. And as the principal strategic arm of the United States, the Air Force would naturally be responsible for giving an account of unidentified flying objects, whether of Soviet or extraterrestrial origin.

Though quick to compete for funding, bureaucracies do not make quick decisions to take themselves in a new direction. Unless faced with a crisis, changing the direction of a bureaucracy requires herculean effort. Making matters worse, bureaucracies operate by developing standard operating procedures (SOPs) to manage most day-to-day affairs. These SOPs make bureaucracies more rational and fair but also far more resistant to change. Once the Air Force got into the UFO business, it was stuck with it. The Air Force tried to remove its obligation for UFO reporting several times—after Sign, after Grudge, and again in Blue Book. But each time they were directed to continue with their reports. Once Air Force researchers determined that UFOs posed no genuine security threat to the United States, the SOPs assumed the role of alleviating public anxiety about UFOs and showing that UFOs were really naturally occurring phenomena. By the 1960s, the Air Force desperately wanted out of the UFO business, and hence it contracted with the University of Colorado.

Finally, bureaucrats desire promotion. People employed in bureaucracies tend to follow orders and not make trouble even if they do not always accept the rationale for a decision. More importantly, however, people who want to get promoted seek those jobs deemed most beneficial for the mission of the bureaucracy. In the Air Force, piloting jet aircraft provides the fastest way to the top. Working on UFO sightings leads to obscurity and insignificance. Thus, much to the dismay of UFO researchers, the Air Force could never be budged from assigning a low priority to monitoring UFO sightings. Without a real threat from the phenomena, the work became a nuisance. Even the relatively low rank of officers assigned to lead the various UFO projects indicates the lack of official interest in the phenomena. But Project Blue Book isn't the only bone of contention between the Air Force and UFO enthusiasts. Even more contentious for many today is a mysterious facility in Nevada.

# AREA 51

ALMOST EVERYONE INVOLVED in promoting Area 51 stories sticks to his or her own favorite interpretation of events and disagrees with the others. This officially unacknowledged military base, located about 100 miles north of Las Vegas, is administered by Nellis Air Force Base. "Officially unacknowledged" means that the Air Force neither confirms nor denies that this base exists. Area 51 sits on a dry lake (Groom Lake) within a geographic area the size of Connecticut. The base was built in order to provide a landing strip for research and development of new, highly classified military aircraft. Products of Lockheed's famous "skunk works" advanced development projects, the U-2, A-12, SR-71 Blackbird, and F-117 Stealth fighter were all developed at Area 51.[28]

All advanced research projects, such as the ones conducted at

Area 51, are "black budget." These development programs are kept secret even in published U.S. government budgets. The programs' actual budgets are kept in a classified annex for review by select members of Congress and their professional staffs. In spite of this, a number of private research groups, such as the Federation of American Scientists, often succeed in detecting black-budget programs.[29]

The reason for this secrecy is simple. While the broad brush strokes of the defense budget may be apparent from published sources, the military needs to keep U.S. adversaries from knowing exactly where its research and development dollars are going.

The history of military competition among sovereign states has always been between the dominance of offense over defense, one weapon over another. For each new weapon, a countermeasure is developed.[30] The advent of stealth aircraft such as the F-117 fighter and the B-2 bomber compels other nations to find ways to detect them with new radar technologies.

During the development stage of military weaponry, secrets are extremely vulnerable to espionage. Any large-scale development project, such as a new aircraft system, involves an enormous number of people and plans—any of which can be compromised by enemy spies. The F-117 and B-2 were specifically designed to overcome one of the more significant technological advantages held by the Soviet Union during the late 1970s and early 1980s: anti-aircraft radar. Knowledge that the U.S. government was developing stealth technology—and especially how—could have led the Soviets to begin countermeasures much sooner in their effort to defeat the U.S. investment.

Nevertheless, the secrecy designed to protect against espionage fuels UFO reports and conflicting interpretations of what is going on. Though some claim to see UFOs in Area 51, the wilder claims of captured aliens and flying saucers have been thoroughly discredited by a number of UFO researchers themselves, more so because of the credibility of the individuals than because of any explanations by the government.[31] Speculation of a high-altitude manned reconnaissance plane (Aurora) being developed at Groom Lake also fuels UFO reports. While the research is likely being conducted (the U.S. needs to continue developing newer aircraft technologies), reasonable doubts exist that anything as spectacular as hypersonic aircraft research is taking place.[32] More in line with the potential threats in the post–Cold War era, Groom Lake is probably the test bed for a number of smaller technological innovations for combating "low-intensity conflict"—terrorism and guerrilla warfare.[33] Such innovations might easily include remotely piloted unmanned aerial vehicles and other reconnaissance platforms.

Black-budget projects are still critical to the security of the United States in the post–Cold War era. During the late 1970s and into the 1980s, the Soviets mounted numerous intelligence operations to steal U.S. research and development (R&D) secrets and reengineer them for the Soviet military.[34] The Blackjack bomber, for one, is an example of how the Soviets saved billions of R&D rubles by reengineering stolen B-1 plans. In the 1990s, the People's Republic of China (PRC) took over where the Soviets left off.[35] The PRC began targeting specific high-end military technology for theft—precisely the kinds of technology being researched at Groom Lake. At this time, the PRC shows no signs of slowing down in its endeavors. As black-budget programs continue at Groom Lake, so too will sightings of mysterious craft.

Area 51, along with Roswell and Project Blue Book, provides a variety of speculative topics for UFO conspiracy theorists regarding U.S. government deception. Natural events and the need for national security offer reasonable explanations behind appearances of government cover-up. But how do these events and speculations evolve into conspiracy theories? The next chapter addresses conspiracy theories and what lies behind them.

CHAPTER 8

# GOVERNMENT CONSPIRACIES

*Mark Clark*

Ever since sighting a UFO as a young man, Jack has spent all his spare time crisscrossing the country to talk with people who report having seen flying saucers or coming into contact with extraterrestrials. What he has heard has been enough to confirm his belief in the reality of UFOs, but definitive proof always seems just beyond reach. That's why Jack has become convinced that the federal government is thwarting his and others' efforts to demonstrate the truth about UFOs.

One day, after speaking on a small-town radio station in a western state, Jack got a call from a man who refused to give his name. They were to meet at a diner, and Jack could recognize him by the man's large green ring. Over coffee and pie in a booth at the back, the man told Jack that he was a highly placed military officer who was part of a counterconspiracy group that intends to expose the government's cover-up of UFOs. The man promised to get in touch with Jack soon and give him incontrovertible proof that he could share with the world.

Jack left the diner feeling thrilled. *It's finally all going to come out!* he thought. *And I'm going to be the one to reveal it to the world!*

Or was a hoaxer laughing at him at that very moment?

CONSPIRACY THEORIES propose that some group of people is working behind the scenes, secretly orchestrating situations for their own benefit and the harm of others. Well organized, powerful, and resourceful, the cabal victimizes weaker people. That's the viewpoint whether the conspiracy theory is one about Jewish domination of Europe (as viewed from Berlin during the Third Reich), about the United States conspiring to spoil the Soviet "worker's paradise" (as viewed from Moscow during the Cold War), or about the U.S. government secretly covering up the true nature of UFO phenomena (as alleged by many UFO buffs).

Furthermore, conspiracy theorists allege that absence of proof for the conspiracy is itself proof, because it shows how well developed the conspiracy is. So, for example, when radical Islamists in Iran accuse the U.S. of manipulating the economies and politics of the Muslim world, they present no evidence of this because their followers require no proof to believe the theory. In a similar way, the absence of physical proof, indeed even of solid circumstantial evidence, indicates to the UFO conspiracy theorist just how secretive and manipulative the U.S. government is when it comes to alien visitation.

Conspiracy theorists never seriously consider the possibility that they may have misinterpreted the facts. Their assumptions rule their judgment. And so their theories do not allow for the sort of self-correction that the scientific method permits. UFO conspiracy theorists, for example, continue to claim that government leaders are covering up evidence of extraterrestrials, regardless of the contradictory evidence that has been amassed.

To be sure, real conspiracies do exist. Vladimir Lenin, along with a small coterie of like-minded followers, conspired to overthrow autocratic Russia. Adolph Hitler and Joseph Stalin colluded to share in the conquest of Eastern Europe. Followers of the Ayatollah Khomeini conspired to overthrow the shah of Iran. Osama bin Laden's al-Qaeda terrorist group secretly planned to attack the centers of American power.

The great majority of real conspiracies, though, are discovered or compromised. Because someone on the inside of the conspiracy compromises the plan, or because others on the outside discover it, few conspiracies can be kept secret for long. Lenin wrote his ideas down on paper to direct his party activists. Stalin and Hitler signed a protocol to their nonaggression pact. The Iranian revolution eventually erupted in violence. And the world watched with horror the devastation wrought by terrorists on September 11, 2001.

It would seem that if the U.S. government were conspiring to cover up evidence of UFOs for the past fifty years, the conspiracy would have come to light by now. Yet UFO believers still raise Roswell, Project Blue Book, Area 51, and other pet subjects as evidence that the government is hiding crucial information from people. In a loud voice they present their provocative theme of the government's involvement with aliens.

This chapter examines how a culture of distrust toward government and a psychological dynamic known as "theory-driven thinking" have helped to create the UFO conspiracy theory. It goes on to suggest evidence that tends to contradict the idea of a government conspiracy. Then it shows how this theory entered mainstream thinking among American citizens. Finally it offers a challenge to conspiracy buffs to apply "Ockham's razor" to their theory.

## POLITICAL CULTURE

THE UNITED STATES has a history of anti-governmental conspiracy theories. An example from recent decades is the belief that members of the government were behind the assassination of John F. Kennedy. Simply put, Americans tend to distrust their government. They are quick to imagine that people in positions of power are working behind the scenes to defraud them in some way.

This distrust on the part of the American people may not be due merely to their suspicious nature; it may have something to do with their constitutional system. The framers of the Constitution deliberately structured the government as a struggle for power. Often mislabeled as a "separation of powers," the arrangement is more accurately described as a "separation of institutions sharing power." The framers understood the human tendency to grab for power, and so they deliberately created separate institutions struggling against one another over shared powers to prevent any one branch—or person—from taking too much power.

Intentionally built into American political institutions, this struggle for power laid the foundation for suspicion of government in the political culture of the United States. Many citizens, especially journalists, interpret this situation to mean that they should distrust government officials.

But the existence of conspiracy theories is more than simply a function of American political culture. After all, conspiracy theories abound in all societies and throughout history. They are not a uniquely American experience (although some evidence suggests that UFO theories were far more common in the late twentieth-century United States than in the rest of the world).[1] The ubiquity of conspiracy theories has something to do with the way humans think.

## THE PSYCHOLOGICAL DYNAMIC

RESEARCH PSYCHOLOGY develops different understandings for how humans think and why they may be prone to conspiratorial thinking. One important set of models, for the study of UFO phenomena, is known as "cognitive psychological models." These focus on the decision rules, shortcuts, and simplifications that people employ to process information.

Psychologists point out that people process information in either a theory-driven (assumption-based) manner or in a data-driven (fact-based) manner. As explained by one scholar from the University of Toronto:

> Theory-driven information processing is a method of interpreting data on the basis of one's prior beliefs or prior knowledge. One begins with an assumption and interprets new information through the prism of that assumption. It is a top-down, concept-driven activity. Data-driven information processing, on the other hand, is bottom-up. One attends first to the data and then develops theories or conclusions on the basis of this raw information.[2]

Because it is inductive, data-driven thinking is time-consuming and mentally taxing. Thus psychologists point out that people engage in theory-driven thinking far more frequently. Because no substantive body of research deals exclusively with conspiratorial thinking, exploring the ideas behind theory-driven thinking may help explain why and how conspiracy theories develop and how they may be tested for rationality.

Scholar Robert Jervis explains five primary and several secondary ideas behind human information processing that lead to theory-driven thinking.[3] First, people tend to construct their own image of reality. Rather than behaving like a camera, accurately recording all visual perceptions in detail, the mind functions analogously, similar to the work of a paleontologist who constructs an image of what an animal looked like based on the unearthed fragments of bones and teeth.

Second, people acquire a set of beliefs and images about their physical and social environment as they mature. These beliefs and images help them make sense of the world around them. In fact, seeing would be impossible without some preexisting beliefs. People who gain their sight after having been blind from birth do not immediately understand the visual world in which they live.

Third, this construction of a person's experience is highly selective. Were it not, people would be overwhelmed by sense impressions. But the filtering process usually takes place outside of their conscious awareness. While reading this book, readers ignore tens, if not hundreds, of other sense impressions: the ticking of the clock, the barking of the neighbor's dog, the noise of cars driving through the neighborhood, the feel of skin next to clothes, the rumble of food in the stomach, and the like. There is an inherent physiological limit on one's cognitive capacities.

Fourth, people tend to see what they expect to see, based on their experiences, on prior advice and instructions, or on their culture. Jervis calls the things people expect to see their "perceptual predispositions." This phenomenon explains why it is so difficult to find typographical errors in one's own writing. People automatically imagine seeing the right letters as they read because they expect to see words, not strange combinations of letters that make no sense.[4]

Fifth, people assimilate information according to their preexisting beliefs. They take in information in such a way that it fits into an existing storehouse of knowledge. Before people make up their minds on a subject, they tend to be open to conflicting opinions and alternative explanations. Once they form an opinion, however, they find it very difficult to change that opinion.

In addition to these five major points, people also seek to maintain cognitive consistency. They attempt to maintain a balance between their beliefs and their perceptions. Seeking additional learning and revising beliefs is one way to maintain this balance. But another, more common way twists, distorts, and interprets (or ignores) data to make the facts consistent with preexisting beliefs.

Jervis points out, though, that one should not be too quick to condemn this biased interpretation of information. If people adjusted their beliefs every time they received contradictory bits of information, they could not act rationally or consistently. When a friend or spouse acts grouchily, people do not immediately assume that he or she hates them, but rather they think that the grouchy behavior is the result of a bad mood or some other explainable cause.

Under routine conditions, such theory-driven thinking is efficient. It conserves mental energy on tasks that are largely mundane or routine. For most things in life, theory-driven thinking is more than adequate to the task of living.

Theory-driven thinking, however, can lead to errors. Because of perceptual predispositions, people tend to look for information that confirms their preexisting beliefs. In fact, they often seek out information exhibiting a "hypothesis-confirming bias" (another of Jervis's terms). An inclination to search for information and evidence that supports and affirms assumptions keeps people from looking for data that might force them to revise their views. Conversely, people likely ignore information that challenges their expectations. If they do seriously consider any discrepant information, they tend to dismiss it as irrelevant or question the credibility of the source.

Interpreting ambiguous information to support preexisting beliefs presents another problem, called "priming." A study demonstrates a classic case of priming's effects. Two groups of participants were asked to evaluate the behavior of people shooting rapids in a canoe. Before seeing the slides, the first group was given subliminal words such as "brave" and "adventurous," while the second group was given subliminal words such as "reckless" and "dangerous." People in the former group were more likely to interpret shooting the rapids in a positive way, assuming the rafters were having fun. People in the latter group were more likely to see the behavior in a negative way, as irresponsible.[5]

This study of biases demonstrates two important results in regard to UFO conspiracy theories. First, theory-driven thinking hinders, if not undermines, people's ability to value alternative interpretations—and theories—of a given event or behavior. People look first to their assumptions and only afterward to the data, and they interpret events in a way that fits their expectations. Second, and equally important, because people seek out information that confirms their assumptions and ignore information that challenges the assumptions, their beliefs resist change under normal circumstances.

Jervis notes three criteria for judging when people are engaging in irrational cognitive consistency—that is, when they are trying to maintain their cognitive balance in an irrational way. People are maintaining cognitive consistency irrationally when they (1) fail to take into consideration a large amount of information that contradicts their views; (2) fail to notice obviously important events that warrant attention; and (3) fail to look for evidence that is clearly available. UFO conspiracy theorists generally meet at least one or two of these criteria.

## Contrary Evidence

CREDIBLE INFORMATION EXISTS that contradicts the UFO conspiracy theory. For one thing, the alleged conspiracy has never been leaked to the public by an insider. For another, no one has ever produced independent proof of UFOs that would expose the government's "lies." And for a third, claims of reverse engineering from alien technology don't hold up under scrutiny. There may be other kinds of contradictory evidence in existence as well, but these three areas are enough to show how some UFO believers have become victims of their own theory-driven thinking.

UFO conspiracy theorists never explain how the U.S. government or a clique of rogue officials within it, could keep a tight lid on an operation of this scale.[6] Several generations of military personnel (from privates to generals), airmen, sailors, scientists, engineers, bureaucrats, elected officials, private security guards, and civilian contractors would have some knowledge or contact with such a secret. And many of them would have an interest in exposing it. If several U.S. officials caught spying for the Soviet Union during the Cold War could profit from the royalties on books about their betrayal (and they did), then it seems likely that at least one disgruntled UFO conspirator might try to do the same.

Keeping secrets in modern-day government is more difficult than ever. On a daily basis, dribbles of classified information appear in such media outlets as the *Washington Times*, the *Washington Post*, the *New York Times*, and the respected magazine *Aviation Week & Space Technology*.[7] Many officials leak information to serve their personal or official agendas; others reveal information to oppose someone else. Yet no credible evidence has ever emerged that U.S. leaders conspire to keep information about UFOs secret.

While administrations in power usually try to stem the steady hemorrhage of classified information, sometimes their efforts fail.[8] If anything, UFO researchers benefit from the government's inability to keep secrets. Using the Freedom of Information Act, many successfully dig up official papers documenting the U.S. government's interest in UFO phenomena.[9] Yet these documents show that U.S. officials were still keenly interested in capturing a crashed saucer at least one year *after* they had supposedly first acquired one.[10]

The UFO conspiracy theorists, however, face an even greater problem of credibility. They have never produced a fragment of an alien spaceship, let alone a whole crashed saucer. None has ever produced an alien body for public display, except in a faked photo on the cover of certain sensational weeklies. Is one to believe that the government has gotten its hands on all of the prime pieces of evidence and then whisked them into hiding before anyone else could secure them—and that it has held each piece safely in its keeping ever since? Apparently so, according to the conspiracy theorists.

## Government Conspiracies    99

If the theorists were to come up with some definite proof of extraterrestrials on their own, it would render the government's supposed conspiracy irrelevant.

And then there's the claim that recent technological advances, such as stealth aircraft or fiber optics, were reengineered from an extraterrestrial vehicle. *Those are stunning advances. How else to explain them?* wonder the conspiracy theorists.

This argument reveals a profound misunderstanding of technological development and indicates poor scholarship on the part of the conspiracy theorists. The truth is that all of the advances cited by theorists as reverse engineering from alien technology are evolutionary developments, not revolutionary ones. Stealth aircraft are an example.

Aircraft engineers have been aware of the radar detection problem since World War II. In fact, during the war, they considered using wooden aircraft frames to avoid radar, but they found that metals were structurally more sound for carrying heavy payloads of men and munitions. During the Cold War, the mainstay of the U.S. bomber fleet was the B-52. When developed in the late 1970s, the B-1 bomber had one-seventh of the radar cross section of the FB-111 (an earlier jet fighter-bomber) and one one-hundredth of the cross section of the B-52. The B-2 Stealth bomber reduced that cross section further, although by how much remains classified.

The innovations for stealth aircraft were numerous but incremental. One innovation applied new ways to hide the engine—one of the biggest sources for radar detection. Another found new types of oblative (flattening) coatings. New designs tried to deflect radar away from its source. In the case of the F-117 fighter, the designers made numerous small angles avoiding 90 degrees (making it boxy-looking), which helped deflect radar. The B-2 bomber designers, on the other hand, used smoother, more rounded surfaces. All this proves non-alien design. If extraterrestrial technology initiated research into stealth, aircraft engineers would have chosen only one method to deflect radar.

To those not paying attention to these technological innovations, such changes might appear revolutionary. Likewise, for those people living at the end of World War II who did not participate in the developing atomic technology, the introduction of the atomic bomb seemed revolutionary. But to all those physicists and engineers who worked long, hard hours during the Manhattan Project, the A-bomb emerged from incremental changes. The U.S. government expended an enormous financial, scientific, and engineering effort to make the atomic bomb—a device that was once only a theoretical possibility. The government did the same thing with stealth aircraft, and it does the same thing with other military weapons today, without ever taking apart a single alien gadget.

Contrary to solid reasoning based on factual evidence, UFO conspiracy theorists often present spurious research and wild speculation based on questionable sources. The retelling of odd circumstances, second-and third hand hearsay, and testimony frequently elicited from strange characters with suspect truthfulness provide most of the "evidence" for the theory. And yet, because the theory is repeated often in books, television specials, movies, and now the Internet, many people believe it.

## POPULAR OPINION

THE UFO CONSPIRACY theory has struck a chord in America. In a 1996 Gallup Poll, about 70 percent of Americans said they believed that the U.S. government knows more than it's saying about UFO phenomena.[11] It should be noted that this poll does not imply that most Americans believe in the reality of UFOs but rather that they believe the government is hiding something. Still, it's striking how many are willing to believe in a government conspiracy.

It's not hard to understand why the conspiracy theory hits home. After Vietnam, Watergate, and the CIA revelations in the 1970s, after the Iran-Contra hearings in the 1980s, and after the impeachment and questionable electoral processes in the 1990s and year 2000, many Americans have become more skeptical of the U.S. government. Baby Boomers and Gen Xers grew up to view their world critically, if not cynically. Even the originator of the popular television series *The X-Files* admitted that his "paranoia and mistrust of authority came of age during Watergate."[12] In popular opinion, the breaches of trust by the U.S. government during the past forty years manufacture strong "primers" for the current wave of disbelief in the trustworthiness of the U.S. government.

But an even more fundamental reason exists to explain why the conspiracy thesis strikes such a chord. The conspiracy thesis about UFOs places any reasonable person, or responsible public official, in a quandary. The quandary arises from the fact that it is impossible to disprove a negative—that is, "The government is not telling us everything" cannot be refuted!

There can be no perfect disproof of the conspiracy theory for a number of reasons. According to the conspiracy theorist, there will always be one more document, one more hangar, or one more person left inside the bureaucracy who has the key to unlock all the secrets of the conspiracy. Further, showing the flaws in the theory, or questioning the credibility of the so-called evidence, only lands a person in the camp of the conspirators—he or she becomes part of the "cover-up." Even UFO conspiracy theorists who argue with each other over specific interpretations of particular events accuse one another of infiltrating the movement and discrediting the real nature of the UFO problem.[13]

No dispute is possible with the conspiracy theory, and ultimately with each conspiracy theorist, for good reason. The conspiracy theory is more of an ideology than a true theory. An empirical (fact-based) theory starts with a hypothesis of causal relationship—X causes Y under such and such circumstances. The researcher seeks to confirm it with evidence. In the natural sciences, laboratory experiments are conducted and checked by others to verify the hypothesis. In the social sciences, a hypothesis is tested against historical evidence. If the facts fail to confirm the explanation, adjustments must be made to the assumptions and hypotheses of the theory (that is, the theory must be adjusted by the theorist). The same must be true for rational cognitive consistency regarding UFOs. People must adjust their beliefs if, over time, the facts do not support them.

Yet an ideologue is under no such compulsion to adjust his ideology. When the facts do not confirm his expectation, the ideologue reinterprets the evidence. Karl Marx predicted communism would emerge in the most advanced capitalist countries, which in the late nineteenth century were Britain and Germany. When communism didn't emerge in Britain, Germany, or even America, Vladimir Lenin came along to explain why it didn't happen. He argued that capitalism developed into imperialism as a mechanism to avoid the revolution temporarily. (Of course, he ignored the embarrassment that imperialism has been around throughout human history.) Later, Lenin led the revolution in feudal Russia—a country that Marx had considered least likely for communism. In the same manner, UFO conspiracy theorists consistently reinterpret when, where, and how the U.S. supposedly got involved in the UFO business. Each new government document and each new military report on the U.S. government's noninvolvement with UFOs become grist for the conspiracy mill. And so the public goes on being duped into thinking that its own leaders are hiding crucial information.

## Ockham's Razor

FOR NONIDEOLOGUES, THERE is an alternative approach to the problem—one more in tune with rational cognitive consistency. And no one has to be a scholar to do it. This approach is called "Ockham's razor." Attributed to the English philosopher and theologian William of Ockham, who died about 1350 A.D., this dictum states that the fewest possible assumptions are to be made in explaining a thing. When two (or more) competing theories explain an event, Ockham's razor suggests that the simpler one is probably closer to the truth. Ockham's razor slices away unnecessary assumptions.

In centuries past, empirical observations presented more and more problems for the geocentric (Earth-centered) model of the solar system—to the point where it was replaced with the heliocentric (sun-centered) model. This same pattern applies to the explanation of the U.S. government's interest in UFOs. From the beginning of the issue, a simpler, more accurate interpretation of U.S. governmental involvement has existed. This interpretation doesn't require the multiplication of ad hoc assumptions or excessive paranoia and speculative inferences. Common sense and a healthy respect for the material interests of governmental agencies go a long way to explain the U.S. government's behavior.

The U.S. government is, in fact, conspiring in a cover-up. But what the U.S. government is concealing isn't related to captured extraterrestrial spacecraft or alien beings. It's concealing its military secrets. Both Constitutional responsibility and political necessity require the government to secure the country from enemies and protect it against military threats. The government defends vital interests by classifying for secrecy certain types of information, intelligence collection (both its sources and its methods), military plans (including operations and training), and weapons development. While a number of critics challenge how many secrets the U.S. government keeps, few challenge its need to keep some.

Quite often, U.S. government, military, and scientific officials find themselves in the uncomfortable position of protecting secrets from inquisitive researchers, not just enemy spies. To do so, they often appear to be—and in fact are—hiding something. Obliged by law to protect official secrets, they can be prosecuted for violating that law. Sometimes they may "neither confirm nor deny" the information; at other times they are required to withhold it; and occasionally they are required to deny it—that is, to lie. Lying may not come easily for most of them. Indeed, people should find it reassuring that, no matter how well-trained public officials become, lying is difficult for many of them, even when national interests require them to do so.

This is the one detail that tellers of UFO conspiracy stories might like their readers to remember when they've forgotten everything else. The U.S. government lies. Yet one must also remember that protecting national security sometimes necessitates lying. By not revealing classified details of developmental programs, government officials may give the appearance of a conspiracy. And then the tendency of people to preserve their central beliefs gives continued life to conspiracy theories.[14] It's far easier to attribute the anomalies in one's beliefs to the machinations of a few evil people than it is to confront one's mistaken assumptions and revise them.

For a variety of reasons, UFO conspiracy theorists engage in irrational cognitive consistency. That is, they try to maintain their cognitive balance in an irrational way by ignoring the absence of evidence for their interpretation, failing to provide a plausible theory for their position, or refusing to believe the validity of contrary evidence given by the U.S. government and other researchers. The three most famous cases cited by UFO conspiracy theorists—the 1947 Roswell incident, the Air Force's Project Blue Book, and Area 51—provide good illustrations of how this occurs. Yet the proposed intrigue isn't quite as bizarre as what theorists like to believe. The scientific paradigm presented in the following chapters, however, may expose an intrigue of a far more fascinating nature than the ones addressed by conspiracy theories.

CHAPTER 9

# Nature and Supernature

*Hugh Ross*

Eleanor, a science teacher at a prestigious prep school in New England, has always rejected anything of a supernatural or paranormal nature, including both God and little green men. She's nearing the end of her career now, but she's proud of the fact that over the years she has always kept up with the changes in her field by reading the scientific literature. She's also proud of the fact that she has steered many of her students away from a belief in God.

So she's troubled that many scientists lately seem to be suggesting there must be something beyond natural processes to explain all that exists in the universe. At first she dismissed this trend of thought as crackpot. But the trickle has turned to a flood and she has to admit that the transcendent-origination hypothesis seems to be a logical conclusion based on the latest findings. Still, it doesn't fit with her lifelong way of viewing things.

When a student recently said in class, "I understand how the big bang led to all the stars and planets, but how did the big bang happen?" Eleanor didn't know how to respond.

EVER SINCE publication of Robert Jastrow's landmark volume, *God and the Astronomers*, in 1978, references to the supernatural—and specifically to God and theology—have become commonplace in books by astronomers and physicists. If a person scans the science shelves at a local bookstore, he will find books with titles like *God and the New Physics, The God Particle, God and the Cosmologists, Reading the Mind of God,* and *Through a Universe Darkly.* What's going on here? Is a mass conversion taking place?

No, it's not a mass conversion. Rather, this development arises from research findings. A mountain of evidence compels the conclusion that reality must exist beyond the physical universe. The big bang (which is actually a whole class of tested and increasingly refined models for the origin of the universe) implies a reality—a Being—beyond the boundaries of cosmic space, time, matter, and energy. This implication has crucial bearing on questions about the nature of UFOs, specifically on the evaluation of hypotheses that attempt to make rational sense of residual UFOs, or RUFOs.

## Beyond Space And Time

THROUGHOUT THE twentieth century, astronomers and physicists toiled tirelessly to construct a workable cosmic model that circumvented the beginning to the universe implied by Einstein's equations of general relativity, especially a beginning as recent as 15 billion years ago. They tried every conceivable means to eliminate it, or at least to push it closer to infinitude, for evolutionary processes seemed to demand virtually infinite time. But all their efforts failed.[1]

In a series of papers produced from 1966 to 1970, three British astrophysicists, George Ellis, Stephen Hawking, and Roger Penrose, uncovered a fundamental reason for the failure of such models. These men discovered that no one had ever extended the solutions of Einstein's equations beyond matter and energy to include space and time.[2] So they did the extension and thereby began the development of what are now called the space-time theorems of general relativity.[3]

These breakthrough theorems declare that space and time are not without limit, not absolute entities. Shattering long-held notions of reality, they say that both space and time began when the universe of matter and energy began. What's more, the theorems hold true under all possible physical conditions, given a universe that contains mass, with dynamics accurately described by general relativity.

Such a revolutionary concept could not be accepted immediately. Though no one raised doubts about the reality of mass, some researchers did question the reliability of relativity. So they subjected it to rigorous scrutiny, test after test, year after year. It passed every one.

Today general relativity stands as the most thoroughly tested and firmly established principle of science.[4] Some scientists suggest that it will soon be referred to as law rather than as theory. Recent additional testing of the space-time theorems has led to the discovery that their conclusions remain valid over even broader conditions than their originators initially imagined. Not only do the theorems hold true in a universe governed by classical general relativity; they also hold true in more exotic models, such as those known as the "cosmic inflation" models.[5]

## THE FIRST CAUSE

SCIENCE, OF COURSE, studies cause-and-effect relationships in the physical universe. Einstein demonstrated and acknowledged that the universe itself is an effect—one first manifested approximately 15 billion years ago. Through decades of research, astronomers, physicists, mathematicians, and others have teamed up to discover more about the origin and development of the cosmos. Based on their work, astrophysicists now confidently assert that the universe began with ten space-time dimensions.[6]

Current cosmic models show that these ten space-time dimensions and the matter and energy associated with them came into existence as a "singularity" (or its equivalent)—an infinitely dense, infinitesimal volume. In the first split second of their existence, six dimensions stopped expanding, and the remaining four (the familiar ones: length, width, height, and time) have been unfurling ever since. Scientists differ on various details of the "big bang," as this origin event is known, but the differences grow smaller as research refines the basic model.

By the rules that govern rationality and the entire scientific enterprise, one must ask, "What caused this first physical effect?" The space-time theorems of general relativity dictate that the Cause of the universe exists—and acts—both beyond and unconstrained by anything within the universe, all its matter and energy and space-time dimensions. The Cause can be described, then, as transcendent. The Cause has the capacity to initiate dimensions of space and time, which in turn release matter and energy as they open and expand.

Is this the point, then, at which science ends and faith begins? Does science have anything more to say about the Cause than to establish its logical necessity and its transcendence? Contemporary pluralism—an understandable if not justifiable reaction against humanity's hierarchical and exclusivist abuses—pressures many scientists to answer, "No." Others disagree. At the very least, however, science has crashed through all physical barriers to discover that some reality—supernature—exists beyond nature.

## The Nature of Supernature

"PEOPLE BELIEVE WHATEVER they want to believe," says the cynic. Most philosophers, scientists, and political scientists would agree that people have that right. But of course, the right does not make all beliefs true. Beliefs vary in validity and can be evaluated by how closely and fully they align with established facts. And many people do care about the rational validity of their beliefs. For those who care to investigate, nature holds abundant clues to its supernatural Source, or Cause. And that investigation may hold keys to unlocking some of the mysteries of UFO phenomena.

As astonishing as it may seem, astronomers today have access to telescopes and imaging instruments through which they can observe to the theoretical limits of the entire cosmos. They have captured images of the universe before any galaxies existed (see figure 9.1). One shows the moment long before stars existed, when light first separated from darkness (see figure 9.2)—a moment when the universe was just 0.002 percent of its present age.

In surveying the universe very nearly all the way back to the origin event, researchers observe its various characteristics. These characteristics, in turn, give shape to human understanding of the transcendent Cause behind (or before) the universe. The profile cannot

## Nature and Supernature 109

be complete, of course, because the observers are confined within the universe, but it can be adequate to rule in and rule out certain attributes of the Cause.

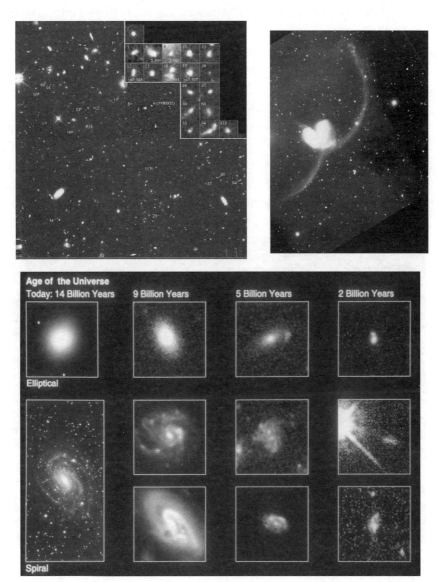

Photo courtesy of R. Windhorst (Arizona State University) and NASA

Fig. 9.1. About 12 billion years ago, star clusters began coming together to birth galaxies. In this particular case, eighteen star clusters are merging together.

First, the measured characteristics of the universe reveal what one might call "fine-tuning" with respect to the needs of living things, from single-celled bacteria to much larger and more intricately complex organisms, such as humans. Each of thirty-eight characteristics (to date) falls within a narrowly defined range of values that, if changed slightly one way or another, would render impossible any conceivable physical life, anywhere, anytime in the history of the universe (see appendix C).[7]

For example, unless the density of matter in the universe is fine-tuned to within one part in $10^{60}$, the universe will expand at the wrong rate for stars like the sun and planets like Earth to form. Likewise, the stars needed for physical life to be possible will burn at the wrong rates unless the ratio of the electromagnetic force to the gravitational force is fine-tuned to within one part in $10^{40}$.

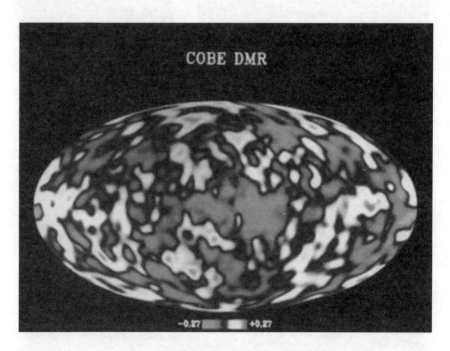

Photo courtesy of Jet Propulsion Laboratory, NASA

Fig. 9.2. Long before stars and galaxies existed, the featureless glow of light from the glow of light from the creation event first separated into regions of light and darkness. This separation of light from darkness occurred when the universe was just 0.002 percent of its present age.

Nature and Supernature 111

The impression of exquisite design goes beyond the gross features of the universe. At least 153 different characteristics of Earth's galaxy and solar system must be fine-tuned to allow for the existence of life on planet Earth (see appendices A and B). Molecules inside the simplest cell show a level of design vastly beyond the best machines humans have built.[8]

This pervasive fine-tuning compellingly suggests such characteristics as intellect, knowledge, creativity, and will, not to mention awesome power. In fact, it suggests purpose and care as well. Such characteristics as one sees in the cosmos belong solely to persons. In other words, the clues point to a Person, rather than to an abstract entity or force, as the designer and manufacturer of the cosmos.

# THE ANTHROPIC PRINCIPLE

THE IMPRESSION OF purpose and care has become so strong as to receive a name: the anthropic principle. The various features of the cosmos point emphatically toward the human being, a complex sentient and spiritual creature, as the beneficiary of transcendent design. The existence of even one potential life site (planet Earth) requires the existence of slightly more than 10 billion trillion stars, no more and no less. (If the observable universe contained fewer than about 10 billion trillion stars, no elements more massive than helium would exist. If it contained more stars, no elements less massive than iron would exist.[9] In either case, life chemistry would be impossible.)

Only a universe about 15 billion years old can produce rocky planets, such as life requires. Only at this age would cosmic conditions accommodate the necessities of a planetary system sufficiently stable for human survival.[10] But this window of opportunity for life does not last long, so life must enter the scene at the right time or not at all.[11]

The planet chosen for life must also be carefully primed and timed also. Hundreds of millions of different species of life must live and die through nearly 4 billion years before human beings can hold any hope of developing a technologically advanced civilization. Making the Earth chemically safe for advanced life, and building up the deposits of metal ores, fossil fuels, topsoil, limestone, marble, gypsum, and phosphates necessary for an advanced civilization requires just as much time.[12]

Within the laws of nature one sees indicators of the Creator's personhood. These laws reflect consideration of, and preparation for, humanity's physical and spiritual needs. These laws show consistency and constancy, without which an intelligent physical being could neither discern them nor trust them. The universe blends order and chaos in exactly the right balance to accommodate the needs of sentient beings. And those sentient beings exist in the precise place that not only accommodates the needs of their existence and survival but also gives them a clear view of the entire cosmos—a rare and perfect place on a just-right planet in a just-right solar system in the just-right spot between the spiral arms of a just-right galaxy to gain a clear view of the just-right universe.[13]

As one of its just-right features, the universe does provide sufficient (though, again, not necessarily complete) information with which to propose, test, and refine human understanding of truth. And the reality of an intersection between nature and supernature leads to the interdimensional hypothesis for UFOs.

CHAPTER 10

# THE INTERDIMENSIONAL HYPOTHESIS

*Hugh Ross*

Vittorio had not wanted to get into UFO research, but because his adviser in graduate school had urged him to do so, he did. And he's stuck with it ever since, doing his conscientious best as a scientist. Now, years later, he finds himself in a difficult position—caught between two camps. In the one camp, there are all the "true believers," as Vittorio calls them, who are sure that little gray men arrive nightly from distant planets. Vittorio's findings long ago led him to reject this "nuts-and-bolts" interpretation of UFOs. In the other camp, there are his peers in the scientific community who want him to denounce all UFO reports as a combination of hoaxes and misperceptions of natural phenomena. But being open-minded, Vittorio has to admit to himself that his research suggests there is something to some of these UFO reports. Where are these undismissable UFO phenomena coming from? Vittorio hasn't admitted it publicly yet, but lately he's kept coming back to a rather spooky third alternative.

DOES SCIENCE inadvertently—and dangerously—open Pandora's box in affirming that nature and supernature interface? The fear that it does drives many scientists and others away from any willingness to discuss religious beliefs in general and paranormal phenomena in particular. Meanwhile, humanity proves persistently religious and incurably curious about the paranormal and supernatural.

Science can and does offer its methodical approach to questions about both tangible and intangible reality. This approach, which incorporates principles of logic, has been applied for centuries in testing truth. It applies to evaluation of the various UFO hypotheses: the natural-explanation hypothesis, the extraterrestrial hypothesis (ETH), the cover-up and conspiracy hypotheses, and the interdimensional hypothesis (IDH). Abundant physical evidence for the reality of residual UFOs (RUFOs) diminishes the plausibility of the natural-explanation hypothesis, which says they are unreal. Abundant physical, historical, and psychological data diminish the plausibility of the cover-up and conspiracy hypotheses, which say they are hidden. Evidence that RUFOs are nonphysical both in origin and behavior diminishes the plausibility of the ETH. And as chapters 3, 4, and 5 demonstrate, the impossibility of interstellar space travel and of a natural explanation both for life's origin and for alternative life sites virtually destroys the plausibility of the ETH. (The question of a supernatural origin for E.T. remains open for the moment—and will be addressed in chapter 15.) But what can scientific methods say about the IDH?

Though at first glance the IDH may appear to be beyond the reach of objective analysis, it can be sifted through the filter of scientific data about the dimensionality of the universe. A brief review may help clarify the point.

In their attempt to develop a model that accurately describes the physical dynamics of the universe, physicists encountered—and eventually solved—a troubling contradiction: Within the boundaries of three space dimensions and one time dimension, either gravity or quantum physics could operate, but not both. Yet proof exists for the operation of both. The short version of the long and exciting story is that a ten-dimensional model (which involves "string theory") works. It accommodates and explains both gravity and quantum physics. As a powerful confirmation, this theoretical model independently corroborates the equations of special and general relativity.

With respect to supernatural reality, these findings do more. They place a lid on Pandora's box. They set up a border around various notions about origins. They provide new testing tools for those people who choose to root their philosophical and religious beliefs in verifiable truth. Thus, they help answer questions about the validity of the interdimensional hypothesis.

## VERSIONS OF THE IDH

PROPONENTS OF THE IDH conclude that RUFOs enter the physical dimensions of the universe from "outside" the four familiar dimensions of length, height, width, and time. But within the general framework of the hypothesis, IDH proponents offer a variety of scenarios. Some suggest that entities appear in this universe from some other physical universe made up of space and time.

Others suggest that these entities are coming from some other dimension that is altogether different from the familiar concepts of space and time, perhaps a spiritual realm. Any unwitting suggestion that RUFOs come from dimensions "between" the space-time dimensions of the cosmos, whether from the six tightly compact dimensions of string theory or from other unidentified dimensions, misconstrues or misrepresents the scientific data.[1]

The most articulate and well-known spokesman for the IDH, French physicist Jacques Vallée, claims that UFO phenomena "represent evidence for other dimensions beyond spacetime." He describes that realm as a "multiverse" all around the cosmos, a realm people have "stubbornly refused to consider … in spite of the evidence available to us for centuries."[2]

While Vallée's hypothesis offers fascinating food for imagination, it is essentially irrelevant from a physics standpoint. General relativity dictates that once physical observers exist in a particular universe, the space-time dimensions of any other hypothetically existing universe could never overlap the dimensions of the universe containing the observers. The observers' universe remains isolated, in the physical sense.[3]

Yet while one must set aside Vallée's concept of a multiverse on scientific grounds, there may be other forms of the IDH that are completely consistent with scientific understanding. The principles of physics do not rule out—and in fact, through the demonstrated reliability of the space-time theorems of general relativity (see chapter 9), they affirm—a reality beyond the cosmos. But to be rational, human speculation about that reality must reflect what the evidence of the observable universe reveals about the Cause and Source of the cosmos.

The characteristics of the universe that make science a doable enterprise—constancy, consistency, order, freedom from contradiction, causality, and so on—provide a reasonable litmus test for any so-called deity or divine revelation. The scientific method can be applied to help determine which, if any, "holy book" or other communication, affirms a reality all around the cosmos consistent with those features.

## The Extradimensional Hypothesis

SCIENCE ATTESTS TO a transcendent cause (or Cause) for the origin of the universe. Most revelatory books, including those supposedly "dictated" via UFOs, allude to trans- or extra-dimensional phenomena and to a transcendent being or force, but these allusions are typically inconsistent and nonverifiable. Most significantly, the God or gods and doctrines proclaimed in these books typically take their shape and limitations from the four large space-time dimensions of the universe: length, width, height, and time.

One book, the Bible, proclaims the existence of a personal Creator who can act independently, outside the cosmos—outside its four large space-time dimensions and six tiny space dimensions. The God of the Bible is in no way restricted to length, width, height, and time or to any of the ten cosmic space-time dimensions, for He is said to be the One who created them.

Of all the holy books of the world's religions, only the Bible unambiguously states that time has a beginning (thus, it is finite, not infinite), that the Creator brought time into existence (thus, caused time), and that God can—and does—exercise cause-and-effect capabilities outside the time dimension of the universe.

> In the beginning God created the heavens and the earth. (Genesis 1:1)

> By faith we understand that the universe was formed at God's command, so that what is seen was not made out of what was visible. (Hebrews 11:3)

The Hebrew phrase translated "the heavens and the earth" refers to the entire physical universe. The Hebrew word for "created" in Genesis 1:1 is defined in the lexicons as "to make something brand-new or to make something out of nothing." Hebrews 11:3 states that the detectable universe was made through that which is undetectable to humans. In other words, the universe was made transcendently; it comes from a Source that exists apart from matter, energy, length, width, height, and time.

Many other Bible passages reflect God's existence "before the beginning of time" (2 Timothy 1:9; Titus 1:2) and identify Him as the Creator of the entire cosmos (John 1:3; Colossians 1:16–17). To place confidence in the truthfulness of such claims requires faith, to be sure, but that faith has its eyes wide open to facts. It is not blind, nor does the Bible endorse blind faith. The Bible does depict "miracles," nonrepeatable events that cannot be directly verified, but only after establishing the frame of reference that would support the possibility of those miracles. That frame, of course, appears in the first story the Bible records—the Creation story passed on and preserved from generation to generation. Because the Creator brought space, time, matter, and energy into existence at will, as that story tells, He certainly possesses the capacity to interact with space, time, matter, and energy in ways that subsequent miracles require.

A New Testament confirmation and clarification of the divine capacity for trans- or extra-dimensional interaction with the cosmos appears in John 20:19. This is the account of Jesus' startling presence and conversation with His disciples following His execution when He appeared in a room that had been locked. The disciples understood the impossibility of a physical body passing through physical barriers. That is why they concluded that the form of Jesus in front of them had to be "ghostly," not physical. But Jesus proved His physical reality by allowing the disciples to touch Him and by eating food in front of them (Luke 24:37–43). Science confirms what people know by experience about the impossibility of passing a physical object through a physical barrier without one or the other breaking. But a physical being (or object) with access to other dimensions (or their equivalent) or with trans- or extra-dimensional capacities could pass through. According to science and mathematics, access to the equivalent of six spatial dimensions would make this phenomenon possible. In the case of Jesus' physical entry through locked doors, He could have rotated the three spatial dimensions of His physicality into supradimensions (or wherever, beyond human dimensions), passed through the barrier, then rotated back into the disciples' three space dimensions.

This doctrine of transcendence holds a pivotal position for two reasons. First, it aligns with observable facts, as attested by rigorous scientific investigation. Second, it requires something more than mere human imagination. Images and ideas, including religious ones concocted by humans manifest the limitations of human visualization and consistency. A revelation and belief system that comes from the Creator can be expected to stretch beyond the limits of human visualization and yet do so without inconsistency or contradiction. The transcendent God and spiritual doctrines unique to biblical revelation, Old and New Testament, speak to the credibility of its contents.

Other characteristics observed in nature by which a person can test the reliability of spiritual communications include intricate, purposeful design of life and careful, thorough provision for the needs of life. The long and growing list of "just-right" conditions in the cosmos, in the solar system, and on Earth speaks volumes about their Source—a Person rather than an impersonal force, a purposeful plan rather than random happenstance, consistent rather than capricious. Such exquisite, meticulous design and abundant provision at all levels for all creatures point to the God revealed in the pages of the Bible, specifically in its many accounts of God's creative work.

No other sacred book communicates so accurately about the natural realm.[4] The mere fact that the Bible correctly describes the personal characteristics of the Creator as manifested in the natural realm is significant. Perhaps more remarkably, the Bible predicts scientific discoveries about the natural realm thousands of years prior to their discoveries. Most reassuringly, the Bible invites and encourages testing, as does no other religious or spiritual communication known to humanity.

## RUFO OPTIONS

THE BIBLE EXTENSIVELY, accurately, and uniquely describes the spirit realm (the realm beyond matter, energy, and the ten space time dimensions associated with matter and energy) on any and all points that can be confirmed by scientific inquiry. Its integrity in these matters lends significant weight to its trustworthiness in addressing those matters that lie beyond the reach of direct scientific confirmation. At the very least, one could argue that because no other holy book or revelation, including the books of the UFO cults, comes close to the Bible's accuracy in depicting physical reality, the Bible stands alone as the best source of trustworthy revelation about the spirit realm.[5]

The Bible declares the existence of God and of two more distinctively spiritual types of creatures: humans and angels. Humans are physical beings with spiritual awareness and spiritual capabilities. Angels, on the other hand, are spiritual beings who are not bound to Earth but are capable of manifesting themselves—even physically—on Earth within limits established by God. According to the Bible, the majority of these creatures remain obedient to God, faithful in His service. A sizable minority, however, rebelled against God. The first and most powerful of these rebellious angels is called Lucifer, or Satan (among other designations). Those who followed him in rebellion are most often called evil spirits, fallen angels, or demons (Matthew 25:41; Luke 10:18; 2 Peter 2:4; Jude 6; Revelation 20:2).

Because human beings remain physically restricted to the dimensions of the cosmos, they cannot account for the unexplained, or RUFO, phenomena, though they can certainly be responsible for some of the UFOs that become IFOs (identified flying objects), and at least some RUFOs may be explained as events in their minds. Angels or fallen angels, however, remain as possible links.

Angels, according to the Bible, are intelligent beings existing beyond the space-time dimensions of the universe, subject to God's spiritual laws but not to all of Earth's natural laws. Angels within certain limits set by God have the capacity to make contact with humans and even to influence their actions. They can take on what appears to be human form and even consume physical food, as in the case of Abraham and Lot's visitors (Genesis 18:8; 19:3). Angels in apparent human form also heralded Samson's birth and greeted the women at Jesus' empty tomb (Judges 13:3-5, 9-21; Luke 24:4-7). Angels do possess some supernatural powers: the two who visited Sodom struck Lot's attackers with blindness, and the angel who visited Manoah and his wife ascended in a flame. But angels, as God's creation, operate within limits set by God, and their powers in no way approach the magnitude of God's power.

The Bible defines the boundaries of God's exercise of supernatural power. While no expression of power lies beyond His capacity, God's actions never violate His character. As Jesus clarified during His years on Earth, God never performs a miracle to dazzle, fascinate, or impress anyone. The miracles He shows to humans are designed to engender humility, to turn people away from exalting anything or anyone above God. Moreover, God does not waste miracles. He apparently performs only those necessary to achieve His purposes.

These biblical parameters, then, offer a basis for making determinations about the source of various RUFO phenomena. If they are from God via His servants, their impact will reflect God's revealed character and values. All creatures who serve Him, the Bible says, will be recognizable by their exaltation of Jesus Christ as the way, the truth, and the life given by God. They will further the fulfillment of God's stated purposes for human beings: to love and serve Him, to love and serve other human beings, and to take care of the world He created. They will never turn attention or worship away from Him toward any other person or being or message of salvation.

If, on the other hand, RUFOs result from the stated purposes and activities of fallen angels, they will reflect a penchant for rejecting or distorting God's authority and purposes. They will draw attention away from God and the historic Christian gospel and turn it elsewhere. They may appear beautiful, dazzling, and inviting ("angel of light"; see 2 Corinthians 11:14) or dark, ominous, and confusing. This impact may be immediate and temporary or gradual and long-term. Given the stark contrast between these two possibilities, angels (ministering spirits) and fallen angels (evil spirits), any RUFO impact that proves truly benign may be considered as having a yet-unexplained but theoretically identifiable source—some explanation other than spiritual, or supradimensional.

With these criteria in mind, one can proceed to look at how RUFOs actually behave.

CHAPTER 11

# A CLOSER LOOK AT RUFOS

*Hugh Ross*

Snivel (also known as Specialist Devil 218) came hurrying in to the office of Master Devil Grimgee. It didn't do to keep the master devil waiting.

"It's about time, 218. We've got another project for you," began Grimgee.

"The usual?" asked Snivel. "You want me to show up as a saucer with flashing lights and make someone think I took them on board?"

"No. That scenario's getting dated. Come up with something new. Like that time you appeared over ancient China as a sky dragon. Just don't come back until you've scared the living daylights out of your prey. I want nightmares. I want virtual paralysis from terror. Got it?"

"Yes, master," groveled Snivel, bowing his way out of the office.

MANY REPUTABLE research scientists and other scholars have done extensive study on both the quantity and distinguishing features of UFO residuals.[1] As Jacques Vallée points out, the problem is not one of inadequate data, given that about a million people per year experience UFO events and that the total number of officially documented residual UFOs (RUFOs) exceeds one hundred thousand.[2] Rather than to accumulate more data, the investigator's task is to establish and categorize the characteristics of those RUFOs that have already been documented.

Do residual UFOs exist? The evidence unequivocally says yes.

The real question is, what are they? While chapter 6 introduced the topic of RUFOs, this chapter gets to the heart of the question about their nature.

## RUFO Characteristics

RESIDUAL UFOS are both real and nonphysical, and as such, they manifest specific characteristics. Examining these characteristics leaves the distinct impression that they have an intelligence and a strategic purpose behind them.

*RUFOs favor certain times and locales.* If neither intelligence nor purpose lies behind the residual UFO phenomenon, then RUFO sightings should correlate directly with the number of potential human observers. The facts demonstrate the reverse correlation. As stated in chapter 6, relative to the number of potential observers, ten times as many sightings occur at 3:00 A.M. (a time when few people are out) as at either 6:00 A.M. or 8:00 P.M. (times when many people are outside in the dark). Furthermore, many more RUFO sightings occur in remote areas than in populated regions.

*RUFOs keep pace with human technology and science fiction.* Scientists who undertake a serious study of the residual UFO database have noted that RUFOs adapt to the culture, technology, and historical context of the human witnesses. Throughout the twentieth century—a time of rapid advance in human technology—residual UFOs reflected parallel technological advances. The same could be said of the whole history of UFOs.

In 1896 and 1897, during a major wave of sightings in many different locations, witnesses described UFOs as strange dirigibles—cigar-shaped, lighter-than-air machines driven by motors attached to propellers.[3] These UFOs were reported to travel faster than any man-made machine, as fast as 150 miles per hour. It is important to note that while a number of patents were taken out in the late 1890s for lighter-than-air craft, no one built or flew such an airship until 1904. In other words, UFOs of the 1890s were slightly ahead of the human technology of their time.

Fast-forwarding to the next major wave of sightings in 1947,

UFO technology apparently advanced to "flying saucers," sometimes moving in formation at speeds of up to seventeen hundred miles per hour.[4] At the time, jet aircraft existed, but none had yet exceeded the sound barrier (about seven hundred miles per hour).

Major surges of sightings were reported for the years 1952, 1957, 1964–1965, 1967, 1973, and almost every year thereafter. Reports indicate that the apparent physical capabilities of RUFOs stayed just ahead of the technology of the most advanced human craft. RUFOs mimicked the limits of published science fiction available at the time of the sightings.

*RUFOs seem to have always been around.* Human technology, as well as the technology portrayed in science fiction, grew slowly before the Industrial Revolution (prior to 1750), rapidly during the Industrial Revolution (1750–1945), and exponentially after the end of World War II. Could the appearance of UFOs with detailed characteristics resembling advanced flying machines be such a recent phenomenon because human technology only recently catapulted to the level at which flying craft mean something to human observers? If that is the case, one may surmise that residual UFOs have always been around, manifesting themselves in forms consistent with culture and technology.

These notions can be explored by digging back into human history to see if earlier peoples reported similar experiences with nonphysical but real entities. Such a study reveals that people throughout all ages of human history have claimed these encounters.[5]

Entities such as aerial people, tyrants of the air, "cloud ships," zephyrs, and "elementals," reported throughout history, resemble the modern residual UFO phenomenon if one considers the human technology factor. Apart from the particulars of technological advance, little discernible difference can be found between typical modern sightings and typical ancient sightings. In fact, if one eliminates the really close encounters, then no differences remain. Written records of ancient sightings dating back at least three thousand years document the same highly luminous balls and multicolored disks, breaking up and coming together, darting around at velocities and accelerations that defy the laws of physics.[6]

More than three hundred years ago (1691), Scottish theologian and minister Reverend Kirk wrote a book describing in detail the paranormal entities that plagued Scottish farmers. The characteristics he listed are indistinguishable from the characteristics of modern residual UFOs.[7] And that is just one example from history.

*RUFOs match the scientific literacy of their witnesses.* In very close encounters with RUFOs, witnesses often claim to receive "messages." The content of such messages usually is tailored to impress the witnesses, especially through new scientific revelations. This scientific information may seem accurate to the witness, but in fact it is not.

Before the twentieth century, UFO "aliens" claimed the moon as their place of origin. When the impossibility of life on the moon became widely known, such aliens gave a different story. Early in the twentieth century, they claimed Venus as their home planet. By the mid-twentieth century, when the public became aware of the intense heat on Venus, the home base for the ufonauts shifted to Mars, Jupiter, Saturn, and other distant solar system bodies. By the end of the twentieth century, when NASA images made it clear that no bodies within this solar system except Earth could sustain advanced life, UFO visitors began stating that they had come from nearby stars.

The claims of ufonauts seemed credible until scientific advances proved them false. In existing reports, no UFO-derived message has ever proved completely accurate in its astronomical content. Regardless of the fact that ufonauts would flunk a graduate-level astronomy course, however, their astronomy knowledge does appear to keep pace with the astronomical literacy of the general populace.

*RUFOs make repeat visits to certain witnesses and sites.* UFO researchers have noted that residual UFOs habitually return to certain sites and to certain witnesses. Someone who has once seen a residual UFO is much more likely to see another one than is any other person to see his or her first one. Similarly, a site once visited by a residual UFO is much more likely to be visited again than is a site never before visited.

At first researchers considered repeat witnesses as attention seekers or as victims of an overactive imagination (if not mentally ill). However, the repeat witness file now contains numerous credible witnesses, many well-trained in sky phenomena (for example, astronomers, meteorologists, and pilots), who have no obvious desire for publicity or recognition. Apparently, residual UFOs target particular witnesses throughout their lives. Stranger still, this tracking sometimes follows a witness's bloodline. That is, the children, grandchildren, and other close relatives of the witness may be visited as well.

The most repeatedly visited sites, according to published reports, are in Brazil and Spain.[8] It is possible, though, that the repetition has less to do with the site than with the witnesses. If RUFOs tend to track certain individuals and families, and if the people remain in a given area, then of course certain sites will receive return visits.

*RUFOs visit a select few.* Though a million people per year may sight UFOs, residual UFOs seem selective in their visitations. Human observers with high occupational or regional probabilities of sighting random UFOs see them far less frequently than do those with low probability. These statistics bear out the conclusion that RUFO encounters are nonrandom.

In 1977 Stanford astronomy professor Peter Sturrock reported results of a survey taken among members of the American Astronomical Society, the principal professional organization of astronomers. Of the 1,356 respondents (professional astronomers), 62 of them (5 percent) reported witnessing unidentifiable flying objects (RUFOs), and a couple of these respondents had seen more than one.[9] However, there was no correlation with relative observing time on the part of these professional astronomers.

As Sturrock points out, even if none of the 1,255 nonrespondent astronomers ever saw an unidentifiable flying object, that still means that proportionally more professional astronomers (2.4 percent) witnessed residual UFOs than did the general populace (less than 1 percent, assuming that of the one in ten Americans who claim to have seen a UFO, less than 10 percent really have seen a RUFO). Part of the explanation is that astronomers are more likely to report a sighting. More significantly, relative night sky observing time did not explain the higher percentage for professional astronomers.

As a member of the American Astronomical Society, I've had the opportunity to meet a number of professional astronomers who indicated witnessing residual UFOs. (With one exception, I do not know whether these astronomers were part of Sturrock's respondents, because none of the contacts allowed their identities to be revealed.) These RUFO witnesses were not astronomers with the greatest amount of observing time. In fact, the sample indicated a reverse correlation. Astronomers with only a few observation hours per year witnessed RUFOs, whereas astronomers logging more than a thousand hours per year saw nothing.

The reverse correlation noted here demonstrates that something besides observing time determines who sees RUFOs and who does not. The most significant factor appears to be the activities that people pursue. Observations reveal that professional astronomers deeply involved in cultic, occultic, or certain New Age pursuits often see RUFOs, whereas professional astronomers who stay away from such pursuits never encounter RUFOs.

This correlation between one's pursuits and activities and the degree of involvement with residual UFOs does not seem to be limited to professional astronomers; it appears to be a universal principle. For the past twenty years, I've made note of this correlation in several dozen lectures and radio and television programs presented on the subject of UFOs. Each time I did so, I received phone calls and letters from people claiming to be the exception to the correlation. Upon deeper investigation, each of these claims either proved not to be an encounter with a residual UFO (that is, the UFO was identifiable with a human or natural explanation) or the individual or one of his or her close relatives indeed was involved in some kind of cultic, occultic, or New Age pursuit. In rare cases, the connection was a close friend that the individual was trying to help leave such activities.

Many documented cases provide support for this correlation where two or more people are together at a residual UFO event but not all experience the event.[10] For example, four people may be standing side by side looking at the same place in the night sky: two see the residual UFO and experience physical and psychological effects; the other two see nothing and experience nothing.

RUFOs appear to be alive. The universality of residual UFOs in the human context can be expressed in a different way: UFOs change with one's ability to perceive them. UFO researcher Whitley Strieber noted that "the fifteenth century saw the visitors as fairies. The tenth century saw them as sylphs. The Romans saw them as wood-nymphs and sprites."[11] Today, human civilization presents a panoply of cultures and beliefs that constantly interact with one another. As one witness who had observed many different UFOs throughout his life stated, "Whatever cosmology or mythology I was immersed in seemed to be the factor for shaping the context and attendant imagery of my experiences."[12]

The capacity for UFO apparitions to adapt so well to the mindset of the human observer, irrespective of that observer's geographical, historical, cultural, and philosophical perspectives, calls into question the deduction that RUFOs are some kind of physical craft. As countless witnesses state, "They seemed to be reading my mind" or "They knew my emotional state" or "They behaved like they were alive." UFO researcher John Keel wrote in his book *UFOs: Operation Trojan Horse*, "Over and over again, witnesses have told me in hushed tones, 'You know, I don't think that thing I saw was mechanical at all. I got the distinct impression that it was alive.'"[13]

RUFOs arouse disturbing emotions. A common denominator for all close encounters with residual UFOs is the type of emotional response experienced by human witnesses. As J. Allen Hynek reports, almost all witnesses are "at an embarrassing loss for words to describe their UFO experience."[14] No witness ever reports being comforted or reassured by his or her RUFO contact. Rather, encounters strike witnesses with intense fear, distress, and anxiety. Sometimes the stress proves so great that the witnesses tremble uncontrollably and for several hours may be unable to speak or move. Nor do these disturbing emotions fade with the passing of the UFO event. The emotional disturbances often grow much worse in the weeks and months following the event. Fear, distress, and anxiety can develop into hysteria, recurring nightmares, and even insanity.[15]

The disturbing emotions aroused by RUFOs are not limited to humans. Certain mammals—evidently only mammals closely bonded to humans, such as the pet dogs of the witnesses—also experience fear, distress, and anxiety. Typically, these mammals react to the UFO before the human witnesses do, but they do not appear to suffer noticeable long-term psychological consequences.

RUFOs cause bodily and psychological harm. Perhaps the easiest way to explode the popular misconception that residual UFOs are benign is to point out the many examples of UFO witnesses suffering bodily and psychological harm. Symptoms such as nausea, headaches, hair loss, diminished vision, diarrhea, swelling, paralysis, sleep cycle changes, and weight loss are common in close encounters of the first and second kind. Less common, but still frequent, are burns, wounds, and even death. A guiding principle maintains that the closer the encounter in terms of physical proximity, the greater the physical injury suffered.

People typically heal from physical injuries. The damage from psychological injuries can last a lifetime. Exposure to residual UFOs at close range causes many witnesses to experience visions, hallucinations, apparent transportation to different regions of space or time, personality changes, and personality disorders. The trauma may be so extreme that some witnesses commit suicide.

These psychological effects can persist for many years following a UFO encounter. All witnesses of encounters closer than a few tens of meters describe the experience as a mental ordeal that does not quickly go away. The witness's psychological state can be so disturbed as to affect everyone living with him or her. More subtle is the fact that many who have had close contact with a residual UFO adopt new belief systems and new forms of behavior. Even after the witness dies, the trauma of the experience may continue in the lives of that witness's family and friends.

RUFOs deceive their human contacts. Serious UFO researchers note a highly targeted misinformation campaign behind the residual UFO phenomenon. Obviously, RUFOs try mightily to portray themselves as advanced humanoids from a distant planet traveling to Earth in metallic crafts. Less obvious is their attempt to steer witnesses toward a changed philosophy of life. The messages forthcoming from close encounters of the fourth kind launch a two-pronged attack: one against naturalism, the other against orthodox Christianity.

The deception behind the residual UFO phenomenon is most evident in the many UFO religions and cults that have developed over the past hundred years. For example, *The Urantia Book*, a tome supposedly communicated to humans by spirit dictation from "super universe rulers," spends the first two-thirds of its 2,097 pages describing a "universe of universes" that is not subject to space and time and the laws of physics. The last third of this UFO bible denies the full deity of Jesus Christ and humanity's need for salvation from its sinful condition. At the very least, as Jacques Vallée (an agnostic) notes, "Our idea of the church as a social entity working within rational structures is obviously challenged by the claim of a direct communication in modern times with visible beings who seem endowed with supernatural powers."[16]

Many skeptics argue that RUFOs can be completely explained as a set of elaborate hoaxes. In one respect they are correct. Human witnesses and their associates, however, are not the perpetrators. As established in chapter 6, the nature of RUFOs and the physical effects that arise from them are far beyond the capabilities of any human team to duplicate. Rather, witnesses and their associates, including their pets and domesticated mammals, seem to be the victims and unwitting instruments of a hoax perpetrated by superhuman authors.

## THE CULPRITS

AT THIS POINT, describing more characteristics of residual UFOs is unnecessary. It can now be determined who is behind the RUFO experiences. Only one kind of being favors the dead of night and lonely roads. Only one is real but nonphysical, animate, powerful, deceptive, ubiquitous throughout human history, culture, and geography, and bent on wreaking psychological and physical harm. Only one entity selectively approaches those humans involved in cultic, occultic, or New Age activities. It seems apparent that residual UFOs, in one or more ways, must be associated with the activities of demons.

Many other scholars, likewise, have deduced that demons dwell behind residual UFO phenomena. Most research scientists involved with serious study of RUFOs, regardless of religious or philosophical perspective, have either drawn the same conclusion or identified an equivalent cause (for example, malevolent beings from another dimension). Jacques Vallée concludes: "The UFO phenomenon represents evidence for other dimensions beyond space time.... The UFOs are physical manifestations that simply cannot be understood apart from their psychic and symbolic reality. What we see here is not an alien invasion. It is a spiritual system that acts on humans and uses humans."[17]

Astronomer and agnostic J. Allen Hynek states that UFOs cause physical effects "in the same way that a poltergeist can produce very real physical effects."[18] With this psychic connection, Hynek claims, "The [residual UFO] problem essentially is solved; that explains why UFOs can make right angle turns, that explains why they can be dematerialized, why sometimes they are picked up on radar and sometimes not and why they are not detected by our infrared equipment."[19]

Another agnostic astronomer, Paul Davies, notes, "No clear distinction can be drawn between UFO reports and descriptions of religious experiences of, say, the Fatima variety."[20]

John Keel, an agnostic who has spent a lifetime researching UFOs, makes the following observation:

> Demonology is not just another crackpot-ology. It is the ancient and scholarly study of monsters and demons who have seemingly co-existed with man throughout history. Thousands of books have been written on the subject, many of them authored by educated clergymen, scientists and scholars, and uncounted numbers of well-documented demonic events are readily available to every researcher. The manifestations and occurrences described in this imposing literature are similar, if not entirely identical, to the UFO phenomenon itself. Victims of demonomania suffer the same medical and emotional symptoms as the UFO contactees.[21]

Lynn Catoe, a senior bibliographer for the Library of Congress, agrees. In reviewing sixteen hundred books and articles on UFO phenomena, she recognized that "many of the UFO reports now being published in the popular press recount alleged incidents that are strikingly similar to demonic possession and psychic phenomena which have long been known to theologians and parapsychologists."[22]

James McCampbell, Jacques Lemaître, and many other physicists who devote significant time to researching UFOs conclude that residual UFOs must be malevolent manifestations from beyond the space-time dimensions of this universe.

## TESTS FOR THE HYPOTHESIS

THE CONCLUSION THAT demons are behind the residual UFO phenomenon is a testable one. According to the Bible, demons can attack only those individuals who, through their activities, pursuits, beliefs, friendships, and possessions, invite the attacks (Leviticus 17:7; Deuteronomy 32:15-43; Judges 9:22-57; 1 Samuel 15:1-16:23; Psalm 106:36-43; Luke 11:14-26; Acts 13:6-11; 17:12-20; 1 Corinthians 10:18-22; Revelation 9:20-21). All that is necessary to further prove the conclusion of demonic involvements, therefore, is to continue surveying people to ascertain who has encounters with residual UFOs and who does not. If the demonic identification of the RUFO phenomenon is correct, researchers should continue to observe a correlation between the degree of invitations in a person's life to demonic attacks (for example, participation in séances, Ouija games, astrology, spiritualism, witchcraft, palm reading, and psychic reading) and the proximity of their residual UFO encounters. Researchers should also continue to observe residual UFO encounters occurring with the greatest frequency and proximity in those communities and nations that manifest the greatest numbers of people opening themselves up to demonic attacks by their activities, pursuits, beliefs, possessions, and friendships.

One reason why research scientists and others may be reluctant to say specifically that demons exist behind residual UFOs is because such an answer points too directly to a Christian interpretation of the problem. However, the only defense to be found against the evil, deception, and supernatural powers manifested in residual UFOs is in Christianity and the Bible.

A second way to test the conclusion that demons are behind residual UFO phenomena is to go beyond the database for close UFO encounters of the first, second, and third kinds, which have been the subjects of chapters 3 through 11, and examine the database for close encounters of the fourth and fifth kinds. Such an examination is the subject of the following three chapters.

CHAPTER 12

# ABDUCTEES

*Kenneth Samples*

Time stopped for Fern that night on the isolated country road. When she came to in her pickup truck, three hours had vanished, leaving no recollection in her mind of what had happened during the missing time. Months later, though, while under hypnosis, she began to recall images. Short, gray creatures with large heads. Levitation into a hovering spacecraft. A painful physical exam.

Fern's recovered memories remain dreamlike in her mind, but she can't forget them. They trouble her day and night. She often catches herself glancing into the sky, fearful of what she may see there.

CLAIMS OF alien abduction give new meaning to ufologists' descriptive phrase "high strangeness." Abduction is, in fact, by far the most controversial part of UFO phenomena. Abductees offer accounts of being captured by space aliens and taken aboard flying saucers. They say they were subjected to medical examination and interrogation and then returned to the point of capture. Only sometime afterward, and usually through hypnosis, were the abductees able to recall the traumatic event. UFO researchers label such abductions as "close encounters of the fourth kind."

A wide variety of people report abduction.[1] Men and women report them almost equally. The education levels of abductees range from high school to graduate school. Both blue-collar workers and white-collar professionals make such claims. Though the ages of abductees vary, primarily the young report abduction. Folklorist Thomas E. Bullard, perhaps the world's leading authority on abduction, notes: "No obvious telltale characteristic distinguishes abductees from their neighbors."[2]

Abduction reports are both common and dramatic. They have captured the public imagination and provided the theme for many a book, movie, and TV show. For many people, such reports are just a source of fun. But the persistence of these reports, combined with the similarities of some of the details, make many wonder whether there's something to alien abduction. To understand the UFO phenomena as a whole, one must consider alien abduction as a key part of it.

This chapter traces the emergence of the abduction experience, lays out the typical abductee scenario, reviews the descriptions of aliens as reported by abductees, and considers possible explanations of the abduction phenomenon. It seeks to answer the question, "What has really gone on for these abductees?"

## A Short History of Alien Abductions

ACCORDING TO THOMAS E. Bullard, who has studied and cataloged hundreds of abduction reports, claims of abduction by aliens came late to the broader spectrum of UFO phenomena.[3] Until the 1950s and 1960s, UFO-related reports remained mostly confined to cases of unidentified aerial phenomena, with reports of alien abduction being quite rare. Those who did report such experiences were viewed as an extreme part of the UFO-related lunatic fringe.[4] But then two high-profile cases, one in the 1960s and one in the 1970s, helped bring the alien abduction phenomenon into the consciousness of the public.

Barney and Betty Hill claimed that, on September 19, 1961, they not only observed a flying saucer and its alien occupants at close range but were subsequently abducted by these space aliens as they drove their car through the White Mountains of New Hampshire.[5] The Hills maintained that they didn't initially remember the abduction event. Later, bothered by nightmares, by deep seated anxiety about UFOs, and by their inability to account for a two-hour period of time during their drive home from a vacation in Canada, they sought help. Finally, through hypnotherapy, the Hills recalled in detail their terrifying experience. After being captured by the occupants of a flying saucer, they suffered a crude and painful medical examination and interrogation by the alien beings and then observed the inside of the spacecraft, including a star map, before being released.[6] According to UFO researcher Jerome Clark, this most unusual and celebrated abduction report made the Hills "two of the most famous UFO witnesses ever."[7] A book based on the Hills' story, *The Interrupted Journey* by John G. Fuller, became a bestseller and was turned into an NBC-TV movie. With equally strong defenders and detractors, the Hill abduction case remains controversial among ufologists more than forty years after the reported event.

By the mid-1970s, abduction reports were beginning to multiply rapidly.[8] Remarkable among the reports at that time was the abduction case involving a twenty two-year-old Arizona woodcutter named Travis Walton.[9] On November 5, 1975, Walton's coworkers reported his mysterious disappearance after he approached a disk-shaped UFO craft that hovered above the trees while they were working deep in Apache–Sitgreaves National Forest near Snowflake, Arizona. All six members of the crew testified that a bluish-green beam of light coming from the craft struck Walton. The crew fled the scene in fear but later returned, only to discover that Walton was missing. Local authorities searched the Arizona wilderness for five days without finding the young woodcutter. Walton finally turned up dehydrated, confused, and seemingly in shock. He asserted that he'd been abducted by short, gray beings with large heads and terrifying eyes. Critics claimed that the story was an elaborate hoax perpetrated by Walton and his brother, who were suspected UFO enthusiasts. Others insisted that the story was based upon reliable testimony. Regardless, Travis Walton's story and the events surrounding his alleged abduction fueled the controversy over alien abduction. Walton wrote a book about the events entitled *The Walton Experience*, which was subsequently made into a 1993 feature film entitled *Fire in the Sky*.

In the 1980s, following the Hill and Walton cases, reports of alien abduction flourished, perhaps encouraged by the publication of two books by New York ufologist and artist Budd Hopkins. Both *Missing Time* (1981) and *Intruders* (1987) drew popular attention to the phenomenon of abduction. Hopkins talked with hundreds of abductees over the years. He contended in his books and speeches that alien beings of a physical nature actually kidnapped untold numbers of people and left the abductees physically and psychologically terrorized.[10] Hopkins worked to bring the reports of abductees to the attention of physicians and mental health specialists.

Abduction reports continued in the 1990s, meeting great skepticism on the part of most scientists and of the mental health community as a whole. However, over the years, scholars and academics from various disciplines, especially psychology, psychiatry, and sociology, overcame their initial negative reaction and began seriously studying the phenomenon.[11] Meanwhile, respected ufologists, who had initially avoided the abduction phenomenon because of its extraordinary claims, began evaluating its significance in relationship to the broader range of UFO phenomena. The Massachusetts Institute of Technology held an Abduction Study Conference in 1992 in which 150 scholars, professionals, and abductees participated.[12] Whether any objective truth provides a basis for these abduction reports remains questionable, but many scholars and researchers nevertheless deemed them worthy of careful evaluation throughout the 1990s.

Not surprisingly, given its exciting nature, alien abduction has become the most popular and most written-about topic among UFO-related issues.[13] At the beginning of the twenty-first century, the subject of alien abduction shows no signs of going away. And with the information that has been gathered, a typical scenario for abduction can now be described.

## The Abduction Scenario

THOMAS E. BULLARD has created a general model for the abduction phenomenon. He lists eight possible episodes in an abduction event: "capture, examination, conference, tour of the ship, journey or otherworldly journey, theophany, return, and aftermath."[14] Not all abduction reports include all of these episodes; some include only a few of them. A look at each of these episodes in more detail follows.

But first, it's worthwhile to note that abduction reports vary considerably in their details. Each report must be considered on its own merits individually and not be made to fit a preconceived pattern. UFO researcher Jacques Vallée points out that the traditional abduction scenario, as frequently presented in the media, often fails to correspond to the actual research data.

> While the casual viewer or the interested ufologist sees this kind of imagery on the television screen, ... detailed research in the field will bring up many other pieces of information. Many abductions do not involve the well-defined phases described by some authors. Many abductions are not traumatic. Many abductions do not involve medical examinations at all. Many abductions do not involve devices that look like spacecraft.[15]

With Vallée's qualification in mind, one can proceed to paint a composite picture of the abduction phenomenon.[16]

1. *Capture.* Abduction often takes place late at night while a person is at home in bed or traveling in an automobile. The person sees an unusual light, sometimes even a flying saucer, and then may experience a form of paralysis or occasionally a form of temporary blindness. The individual's experience can take on a dreamlike nature, with him or her observing alien beings passing through walls or floating through the air. The aliens take the person on board a spacecraft, and then that person may lose consciousness. Most abductions involve a single captive.[17]

2. *Examination.* The aliens sometimes subject the abductee to what is described as a crude and painful medical examination. The aliens place the abductee on a table and remove his or her clothes. Often they then clean the abductee's body and examine him or her by hand or with a machine. The alien examiners pay special attention to the person's reproductive organs, and at times a man's sperm is taken or a woman's eggs are extracted by inserting a needle into the abdomen. Some abductees have reported that blood, skin, hair, and tissue samples were taken; others have reported that the aliens implanted small objects into their body. On occasion, abductees have reported being subjected to sexual assault. Some have even said that

the abduction was arranged for the purpose of selective breeding.[18]

*3. Conference.* After the physical examination, the aliens interrogate the abductee, usually communicating through telepathy. The person may communicate with several different aliens, including one who is clearly the leader. The aliens may instruct the person concerning an important task, but often communication on the part of the alien beings turns out to be "evasions, lies, or nonsense."[19] On some occasions, the communication is positive—the aliens are kind and reassuring. However, on many other occasions, the aliens seem indifferent or even hostile.

*4. Tour of the ship.* While it is rare, some abductees speak of being given an opportunity to tour the flying saucer and view the engine room or the bridge. Bullard mentions an even more bizarre place of visitation. "Another destination has become popular since the publication of *Intruders* in 1987," he says, "as the captive visits an incubatorium where fetus-like forms float in jars or tanks, or a nursery where cradles contain small, fragile, seemingly sick babies. On other occasions, an abductee may meet an older child, a half human and half-humanoid hybrid."[20]

*5. Otherworldly journey.* In approximately 25 percent of abduction reports, the abductee claims to experience a journey to another world.[21] This journey may involve the spacecraft or may be some kind of visionary experience. Some abductees convey detailed information concerning the other world's environment.

*6. Theophany.* At times, abductees have a religious or spiritual experience in which they claim to encounter the divine. Bullard explains: "On rare occasion an abduction closes with a spiritual experience of some sort. The abductee may hear the voice of God, witness a visionary scene, or participate in a ritual."[22] This topic warrants further elaboration.

According to John Whitmore, the abduction phenomenon is essentially religious, exhibiting "religious overtones or similarities with more traditional types of religious experience."[23] This religious dimension of the abduction experience is seen in abductees encountering the "Other" and in their being recipients of physical healings, receiving spiritual guidance, and experiencing "salvation from above."[24]

Whitmore enumerates five common changes in the religious views of persons following their abduction experience.[25] These persons (1) become more spiritual in their outlook; (2) identify a supernatural force behind the universe; (3) hold religious views that are more syncretistic and pluralistic; (4) develop paranormal talents, such as out-of-body experiences or telepathy; and (5) change permanently their overall world-and-life views.

Many people describing their abduction experiences as traumatic and terrifying may contest Whitmore's assessment. Some testify that their encounter with aliens produced long-term negative effects in body, mind, and soul. The common changes in religious thinking that Whitmore enumerates may in the end be deceptive and counter to ultimate truth.

7. *Return.* Following the examination and conference or whatever has gone on during the abduction, the aliens bid farewell to the abductees, sometimes promising to find them again and often instructing them to forget the event. Then the aliens return the abductees to their homes or automobiles. An abductee may describe feelings of either relief or sorrow at being released.

8. *Aftermath.* Though often having no clear conscious memory of the abduction (supposedly the memory has been wiped away by the aliens), the abductee begins to suspect that he or she has undergone a life-changing event. Alarming factors often provide evidence for this realization.[26] For instance, mysterious markings or scars may appear on an abductee's body, especially on or around the genital organs. Furthermore, an abductee realizes there is time—usually from one to three hours—missing from his or her recollection. Abductees experience fragmented memories of the events and are haunted by horrific recurring dreams about them.

Because of the enduring psychological trauma, the person typically reaches out for professional help, usually to a counselor or therapist familiar with the abduction phenomenon. By use of hypnotic regression, the hypnotist discovers lost or hidden memories of the abduction.

Whether the so-called recollections recovered under hypnosis correspond to objective reality, or whether instead they are the product of subjective influences, is a hotly contested topic among ufologists and the mental health community.[27] According to some in the psychiatric community, under particular circumstances hypnotic regression can prove to be a useful tool in uncovering lost or hidden memories. On the other hand, some think the practice to be highly unreliable for discovering true memories, as the hypnotized person is in a highly suggestive state. Leading or suggestive questions by the hypnotist may inadvertently implant false memories of abduction in the person's mind. Additionally, the hypnotist may actually uncover false memories or forgotten fantasies. As Bullard points out: "Fantastic memories of past and future lives, satanic rituals, and alien abductions have emerged under hypnosis."[28] However, significant numbers of abductees do claim to recall their abduction events apart from hypnosis.[29]

Over the long term, abductees often report experiencing further psychic or paranormal experiences, such as seeing apparitions, experiencing poltergeist phenomena, or developing ESP.[30] They report additional sightings of UFOs and may experience additional abductions. About half of all abductees claim to have been abducted more than once.[31]

Clearly, the encounter with alien abductors, however closely that encounter fits with the typical scenario, is an experience that marks people permanently.

## The Abductors

THE QUESTION EVERYONE most wants answered by abductees is "What were the aliens like?" If the abductees actually spent time on board a flying saucer, as they claim, then they are in a position to relate something about humanity's fellow residents in the universe. This naturally provokes curiosity.

As with the abduction scenario, details about the aliens themselves vary considerably in the reports given by abductees and yet there is enough similarity to draw some general conclusions. While what follows is not intended to be comprehensive, it provides a paradigm for thinking through abductees' reported experiences with aliens. It briefly examines reports of the aliens' appearance, origin, nature, and intention.[32]

*Their appearance.* The majority of abductees describe aliens as appearing in one of two ways, both humanoid.[33] These are known as "the grays" and "the Nordics."

Abductees who encounter "the grays" describe them as short (two and a half to four feet tall), grotesque creatures with gray skin, no hair, disproportionately large heads, lipless mouths, slanting dark eyes, a vestigial nose and ears, a frail build, and hands with fewer than five digits. The majority of abductees, especially those from America, describe aliens of this type.[34] Often described in the hierarchy of aliens as the laborers, "the grays" are usually indifferent or even hostile in their treatment of abductees.

The second group of aliens is known as "the Nordics" because they are tall and blue-eyed with long, blond hair. These "beautiful aliens" take command when the two types of aliens appear together. Common in British abduction reports (though rare in American ones), the Nordics usually have a friendly demeanor.[35]

*Their origin.* Abductees are not always able to tell much about where their abductors came from. But when they report on the origin of the aliens, they say they come from neighboring planets, from distant galaxies, or even from other dimensions of reality (nonphysical or spiritual reality).

*Their nature.* Abduction reports differ sharply as to the true nature of the aliens. Most abductees describe their captors as physical beings that appear humanoid. Others, though, describe them as psychic realities or even as spiritual beings or visitors from other dimensions.[36]

*Their intentions.* Two general views attempt to explain the true intention of the aliens as either benevolent or malevolent.

Some abductees describe kindly aliens who offered them moral and spiritual injunctions to pass on so that humankind could avoid some type of catastrophe or evolve spiritually. Accordingly, these abductees view their abduction as a harmless and indeed a positive experience. Some of them claim that the episode left them mentally, morally, and spiritually uplifted.

Other abductees, however, believe that the aliens' intention is evil. In fact, several prominent ufologists believe that the abductors are attempting to recondition and control people's basic belief systems.[37] Others argue that the aliens' intention is physiological exploitation. Thus, the purpose behind medical experimentation is to extract human genetic material in order to create a hybrid alien-human race.[38]

For what it's worth, that's the picture of aliens that emerges from the reports of abductees. But it leaves the central question: Is the abduction phenomenon fact or fantasy?

## Explanatory Hypotheses

WHAT CAN BE made of abductee claims? Bullard puts the dilemma well when he writes, "Abduction reports combine outstanding information with problematic veracity, the testimony of persons too credible to doubt with descriptions too fantastic to believe."[39]

Some believe that alien abduction is a real, objective, and factual experience. In support of their view, these believers appeal to physical evidence such as scars, stained clothing, and implants found in the body. They also point to similarities among the recollections given by abductees, arguing that these people appear to be telling the same sort of story with matching details.[40] Furthermore, they assert that the vast majority of abductees do not seem to suffer from a mental disorder but rather seem to be sincere, honest, and ordinary folk from various walks of life with no motive to deceive.

Of course there are others who reject the factual nature of alien abduction. These skeptics argue that the phenomenon is solely based upon anecdotal evidence.[41] They say there is little or no corroborating testimony and no undisputed physical evidence (artifacts, debris, implants, and so on) to verify abduction reports. In some cases, they say, abduction reports are due to attention seekers practicing a deception on the public. Further, they point out that even honest people are susceptible to bad judgment, self deception, fantasy, delusions, and hallucinations. The alleged abduction memories could also, they say, be explained as culturally conditioned or as hypnotically implanted. And alien abduction is, in the minds of most skeptics, flatly implausible, unfeasible, and contrary to common experience and established scientific truths.[42]

As these two divergent viewpoints suggest, there are different explanations for alien abduction. A number of hypotheses should be considered, some of which contradict the others and some of which may complement each other.[43] Each hypothesis is identified with at least one popular UFO theorist.

*Brain effect.* Canadian neuropsychologist Michael A. Persinger argues that the abduction experience has no basis in objective reality. Instead it is caused by naturally occurring electrical charges (possibly from earthquakes) that affect the temporal lobe of the brain, thus causing a hallucinatory experience in the individual.[44] A philosophical naturalist, Persinger believes that near-death experiences and religious experiences in general can be explained in a like manner.

*Hoax or fantasy.* UFO skeptic and debunker Philip J. Klass argues that alien abduction has no basis in reality. Rather, each reported abduction is either a hoax or a fantasy. Sometimes people stage a false abduction because they are seeking fame or fortune. At other times, so-called abduction memories are merely fantasies drawn from the subconscious of the individual during hypnosis.[45]

*Dormant memory of birth.* Cornell astronomer and popular science writer Carl Sagan once suggested that memories of alien abduction recovered during hypnosis are actually latent memories of the human birth experience. He saw parallels between the claims of abductees on the one hand and the traumatic experience of natal delivery on the other.

*Dream or hallucination.* University of Kentucky psychologist Robert A. Baker advocates a natural psychological explanation for abduction. He argues that abduction claims may be the result of a particularly vivid dream state, a fantasy-prone personality, or even a hallucinatory experience.[46]

*Psychological cover for abuse.* New York psychiatrist Rima E. Laibow says that a significant number of the abductees she's encountered have been physically or sexually abused as children. Therefore, the abduction phenomenon may be a subconscious cover or screen to hide painful memories of a different kind.[47] Laibow suggests that many people with psychic or paranormal abilities have suffered trauma, hardship, or abuse.

*Transformation of human consciousness.* Whitley Strieber, horror fiction writer and alien abductee, is the author of two books on abduction (*Communion*, 1987; *Transformation*, 1988). He and Harvard psychiatrist John E. Mack (*Abduction*, 1994) have suggested that abduction is an extraterrestrial phenomenon with a benevolent intention. They say that while abduction may involve some physical pain for the abductee, its purpose is to enlighten the abductee's consciousness and thus represents an opportunity for spiritual development.[48]

*Extraterrestrial experimentation.* Abduction specialists Budd Hopkins and David M. Jacobs believe that extraterrestrial beings are not only visiting Earth but are actually abducting millions of human beings from various parts of the world in order to extract human genetic material for the purpose of creating a hybrid race.[49] They say abductees are tagged like animals for alien laboratory experiments. Hopkins and Jacobs believe that these same individuals (sometimes including family members) are abducted again and again over decades to facilitate ongoing research.

*Contact with extradimensional beings.* Longtime ufologists John A. Keel and Jacques Vallée argue that some abduction experiences can be explained in terms of real encounters with malevolent intelligences, psychic or paranormal, from another dimension of reality.[50] While Vallée's views have evolved over the years, both he and Keel have suggested that there is a parallel between UFO intelligences and what Christian theologians refer to as demons.

*Demonic implantation of memories.* While agreeing in large measure with Keel and Vallée, Christian apologist and longtime ufologist John Weldon suggests that some abduction experiences may be explained as demonic implantation of pseudomemories. This implantation can occur when individuals involve themselves with the occult practices that are characteristic of the broader UFO phenomena.[51]

Whichever of these explanatory hypotheses is nearest the truth, it's not surprising that people form irrational beliefs. While people sometimes form their beliefs based on rational factors (facts, evidence, credible authority, and so on), at other times they form convictions based on nonrational factors (emotion, self interest, peer pressure, and so forth). In other words, people sometimes believe what they want to believe. The human mind shows an uncanny ability to create and embrace an intriguing fiction as reality.

Possibly the most detrimental result of humanity's propensity to reject God's authority is the human disposition to suppress truth, especially truth about God, and to choose instead to believe what they intuitively know to be false (Romans 1:25; Ephesians 4:18). People can and do create a counterfeit reality in order to avoid confronting the ultimate reality of God and their moral accountability to Him. And that false reality likely results in some reports of alien abduction—a realm of high strangeness, indeed.

Moving further into the realm of high strangeness in the next chapter, the topic of contactees is considered. While abductees are people who claim to have been taken aboard alien spacecraft, contactees are people who claim an even closer contact with aliens—they claim they are mouthpieces for visitors from distant places. What do they have to say?

CHAPTER 13

# CONTACTEES

*Kenneth Samples*

The suntanned and smiling man up on the stage seems so serene, so wise. He gives the impression of someone who has passed through a life-changing experience that has given him inner peace and a message the rest of us could use. He's talking about Rotran, the being from beyond Alpha Centauri who communicates with him on a regular basis.

"Rotran and his race are loving and kind," says the seminar speaker. "They are wiser than us and, as they have been monitoring our progress, they can tell that we are at a crossroads. We are poised to either descend into a new Dark Age or else expand human consciousness to a higher level. They are prepared to teach us how to live a life of peace and harmony like they have on their planet, if we are willing."

IF REPRESENTATIVES of a superior form of life were visiting Earth, might they not have some insight humanity could use? And if they wanted to share messages with humans, might they not choose to speak through some of the more receptive individuals? That is exactly what is happening, say those dubbed by ufologists as "contactees."

A contactee is someone who claims to experience direct personal contact with alien entities on a regular basis. Contactees assert that UFO occupants have selected them to convey critical information to the human race. Thus, in terms of human-alien connection, the contactee phenomenon goes far beyond even the abductee phenomenon.

Contactee experience differs from abduction experience in four distinct ways. First, while abduction is often a negative experience, the contactee experience is generally positive. Second, whereas an abduction experience is usually a single episode, contactees maintain long-term contact with extraterrestrials. Third, while abductees are primarily the objects of alien research, contactees channel alien messages and revelations to other humans. Fourth, although abductees often undergo changes in their religious beliefs, contactees sometimes go much further in actively forming new religious groups.[1]

Contact is an important part of UFO phenomena and is worthy of study in its own right. In this chapter the connections between the contactee experience and the occult will be traced. Some of the more famous contactees will be met, noting similarities among their experiences and discovering how they describe the aliens they channel for.

# THE OCCULT ANCESTRY OF THE CONTACTEE MOVEMENT

THE UFO CONTACTEE movement emerged at the height of the flying saucer age in the 1950s. That emergence will be examined. But first, one must look even earlier in time for the roots of this movement, because it is actually a continuation of occult beliefs and teachings that coalesced around various groups and individuals over roughly the past two centuries.[2] A few of the more important of these groups and individuals are considered, starting with Emanuel Swedenborg.

Two hundred years before the flying saucer age began, Swedish scientist and mystic Emanuel Swedenborg (1688–1772) wrote *Earths in the Solar World* (1758). In this book he revealed his personal journey to several planets in the solar system and beyond.[3] He claimed to have encountered extraterrestrial beings in his travels and provided specific details concerning their civilization.

Swedenborg's extraterrestrial encounters clearly took place through psychic or occultic means. He never spoke of a flying saucer, because he didn't need one. Religion scholar J. Gordon Melton, who has extensively studied the contactee phenomenon, explains:

Most nineteenth-and early-twentieth-century contact with extraterrestrials occurred in a spiritualist context, more likely than not in a séance. The prime mode of contact was a phenomenon quite familiar to psychic researchers, namely "astral travel." A person experiencing astral travel senses his/her body and consciousness separately, and while the body remains in one place, the consciousness travels around. Thus Swedenborg and [other] mediums ... could go into trances and travel to the various planets.[4]

Another leading occult figure who spoke of encounters with extraterrestrial entities was Madame Helena Petrovna Blavatsky (1831–1891). She founded the influential occult religion Theosophy in 1875.[5] Blavatsky, a former spiritualist (one who attempts contact with the dead), developed a new occult religious system centered upon belief in "ascended masters."[6] These ascended masters make up a hierarchy of supernatural beings placed between humanity and the divine. Proponents of Theosophy believe that the ascended masters try to assist and guide humanity's continuing spiritual evolution.

Included in Blavatsky's list of ascended masters are beings from other planets. Thus Blavatsky claimed to communicate with extraterrestrial beings through occultic means. She included among her ascended masters beings she called "Lords of the Flame" and the "Lord of This World."[7]

An offshoot of the Theosophical Society, the I AM movement was founded by Guy and Edna Ballard in the 1930s.[8] The I AM movement affirmed belief in progressive revelations from the same ascended masters identified by Blavatsky. Guy Ballard, however, made revelations from extraterrestrials a centerpiece in this movement. The I AM movement may be the first religion centered upon revelatory messages from alien entities through occultic means.[9]

Written in the 1920s, *The Urantia Book* purports to convey messages from numerous extraterrestrial or supermortal beings through the process of automatic writing. In this occult practice, a medium records messages from the spiritual realm (though in the case of *The Urantia Book*, the actual human recorders remained anonymous). More than two thousand pages in length, this work claims to reveal previously unknown truths about the cosmic histories of the Earth ("Urantia") and new revelations and interpretations about the life, mission, and teachings of Jesus Christ.[10] With its alleged extraterrestrial messages, *The Urantia Book* remains popular today within the circles of UFO religionists.

Scholars are not in complete agreement as to how the contactee movement should be traced historically, but they do agree that all of its major connections are within occult religion. Leaders in the contactee movement also exhibit a general pattern concerning their extraterrestrial contacts. Three points are worthy of consideration:[11]

1. Alien entities are said to be primarily from neighboring planets, especially Mars and Venus.
2. The alien messages are communicated through psychic experiences, particularly through "mediums," by means of what is popularly referred to as "channeling."[12]
3. When communicating with aliens, the contactee uses telepathy (direct communication from one mind to another).

Individuals and groups who influenced the contactee movement from its earliest days claimed to channel alien beings, but they didn't speak specifically of flying saucers and personal visits from extraterrestrials. The contactees of the 1950s made those audacious assertions.

## CONTROVERSIAL CONTACTEES OF THE 1950S

THE MOST FAMOUS of the 1950s flying saucer contactees, George Adamski (1891–1965) ushered in the "age of the extraterrestrial" by proclaiming that he'd actually met an extraterrestrial in a southern California desert.[13] Adamski, long associated with occultism and the metaphysical movement, wrote fictional stories in the 1930s about beings from other worlds. But now he claimed an encounter with a long-haired man from Venus named Orthon. Regular visits and communications from extraterrestrials followed.[14] Adamski described Orthon in these words:

He was round faced with an extremely high forehead; large, but calm, gray-green eyes, slightly aslant at the outer corners; with slightly higher cheekbones than an Occidental, but not so high as an Indian or an Oriental; a finely chiseled nose, not conspicuously large; and an average size mouth with beautiful white teeth that shone when he smiled or spoke.[15]

Adamski asserted that on numerous occasions he had boarded a flying saucer and traveled to neighboring planets inhabited by benevolent beings called "Space Brothers." His travels included meetings with "Venusians, Martians, and Saturnians," and he even attended a "galactic council."[16] Having taken photographs of an alleged flying saucer within the Earth's atmosphere, Adamski distributed copies of the pictures to the public. His books *Flying Saucers Have Landed* (1953), *Inside the Space Ships* (1955), and *Flying Saucer Farewell* (1961) made Adamski affluent, and in the words of one researcher, he "became a significant international occult celebrity."[17]

The 1950s contactee movement in general, and Adamski in particular, were an embarrassment to serious researchers of unidentified aerial phenomena. Ultimately, Adamski's stories and claims, especially his so-called physical evidence, could not withstand sustained scientific and logical scrutiny, and he was exposed, at least to some degree, as a hoaxer and a fraud.[18]

Despite his being all but discredited in his later years, Adamski's contactee claims inspired many others in the contactee movement. The ensuing occultic religious influence continues into the present. Religion scholars J. Gordon Melton and George M. Eberhart provide this assessment of Adamski and his subsequent influence:

> Sifting out the fraud and fantasy from what George Adamski really believed and experienced may well be impossible. Whether prophet or faker, possibly both, Adamski inspired a number of others who also claimed contact with extraterrestrials. Characteristic of the contactees of the 1950s were encounters facilitated by some means of psychic phenomenon, usually the reception of

telepathic messages over a period of time (what is today called channeling) and a subsequent call for a religious response to content of the channeled messages. Most of the contacts also resulted in the formulation of a spiritual group which resembled either a spiritualist or theosophical organization.[19]

Another controversial contactee of the 1950s was George Van Tassel.[20] With a background in aviation, Van Tassel moved to the California desert, where he claimed to have been contacted by various aliens. Van Tassel said that he was able to channel telepathic messages from the alien commanders of various starships.[21] One of his contacts, named Ashtar, became one of the most widely channeled extraterrestrial intelligences in contactee circles.[22] According to Van Tassel, another alien (named Solganda) allowed him to tour a spacecraft and gave him the monumental task of building a machine that would reverse the aging process and serve as a time machine.[23] Van Tassel's book, *I Rode in a Flying Saucer* (1952), revealed his extraordinary contactee experiences. At his annual Giant Rock Interplanetary Spacecraft Convention, Van Tassel introduced many new contactees to the interested public.

While the contactee movement peaked in popularity at the beginning of the 1960s, the UFO-related occult religious groups that followed this peak recognized an opportunity to stake out their territory. A broader acceptance of Eastern mystical and New Age teachings in the West expanded these groups' ability to establish a connection with people. However, analyses of typical contactee scenarios and of the tenets of UFO-related religious groups reveal how ephemeral and unsubstantiated their belief claims are.

## THE TYPICAL CONTACTEE SCENARIO

CENTRAL TO THE contactee phenomenon is the understanding that space entities desire to communicate with humankind. These alien beings select certain individuals to serve as a point of contact, or channel, between themselves and the human race. They then make contact with these individuals in one of two ways: either face-to-face or psychically.[24]

Some contactees, especially the early ones of the 1950s, claimed to have face-to-face, physical encounters with space aliens. They sometimes mentioned boarding metallic spacecraft and touring the solar system with their new space brothers. On occasion they produced evidence (for example, Adamski's alleged photograph of a spacecraft) to substantiate their extraordinary claims. Yet none of the physical evidence was unimpeachable. In fact, the evidentiary basis for the assertions of the early contactees is sorely lacking.

More typical among contactees than the claim of face-to-face encounters has been the claim of receiving messages from extraterrestrials through psychic or paranormal means.[25] Thus the contactee receives the critical and privileged mission of serving as a channel for alien intelligence from other worlds (or possibly other dimensions of reality). This deeply occult experience centers upon the contactee channeling telepathic messages and revelations from alleged extraterrestrial beings. While in a trance, the contactee may also record written messages (known as automatic writing). Entrusted with "the secrets of the universe," the contactee sometimes believes himself or herself to be the aliens' "envoy on earth."[26]

Some contactees claim to have both physical encounters and psychic experiences. But whatever the means of contact, the contactees present themselves as faithful deliverers of important messages from alien beings.

## THE ALIEN DIMENSION

WHEN CONSIDERING ABDUCTEES in the previous chapter, the image of aliens was looked at in terms of the aliens' appearance, origin, nature, and intention. Contactees' descriptions of aliens tend to be similar, in those categories, to those of the abductees. But there are some differences as well.

*Appearance.* Abductees tend to describe aliens as either "grays" or "Nordics." Contactees may do so as well. They may say aliens appear as humans, as humanoids, as animals, as robots, or as apparitions.

*Origin.* Abductees generally say that aliens are either from distant parts of the universe or are from other dimensions. Contactees, in the early days, would often name Venus or Mars as the home planet of their aliens. But that has changed, no doubt because of the ever-increasing scientific data showing that life cannot exist on other planets in Earth's solar system. Contactees' explanations of the origin of aliens are becoming more like that of abductees.

*Nature.* Many contactees, like many abductees, describe aliens as life-forms from outer space. Yet some contactees describe aliens as formerly physical beings who have evolved to an advanced spiritual state of existence. As researcher John A. Saliba notes, "Many contactees ... write about UFOs and space beings as if these were psychic phenomena, belonging to a different time/space dimension that lies beyond the scope and province of modern science."[27]

*Intention.* Contactees maintain the same two general views as abductees regarding the intention of alien contact with humanity. Some aliens demonstrate a benevolent intent; other aliens manifest a malevolent intent.

Many of the early contactees indicated that the aliens were technologically, morally, and spiritually advanced beings who desired to express their benevolence by either helping humanity avoid potential catastrophes (such as a nuclear holocaust) or guiding humankind through the next step in its evolutionary development.[28] Religion researchers Robert S. Ellwood and Harry B. Partin note the following: "Invariably, these groups believe that many, at least, of these astounding vehicles [UFOs] bear envoys from a superior and benevolent civilization from another world who have come to warn and aid us in our folly."[29] The alien messages and revelations communicated through the contactees often involve moral, metaphysical, and theological injunctions and reflections. The alien's benevolent intent is confirmed in the minds of some contactees by the fact that the contactee experience is usually positive (unlike the abduction experience, which is often described in very negative terms).[30]

Some recent contactees have described aliens that are indifferent or even hostile to humankind. While reported negative contactee experiences appear to be in the minority, UFO researcher John Keel argues that there are many "silent contactees" who have contacts with what Keel calls "ultraterrestrials" (malevolent, paranormal beings).[31] Keel and others believe that these alien intelligences are attempting to manipulate and control the contactees for their own nefarious purposes. There have been reports of contactees suffering extremely harmful effects through their interactions with these alleged space aliens.

It's important to carefully evaluate the contactee experience. Reports of apparent benevolent intent may appeal to some, but these may also lead to involvement in UFO cults—the topic of the next chapter. And what if that which appears to be benevolent should turn out to be something not nearly so innocent?

CHAPTER 14

# UFO Cults

*Kenneth Samples*

As the newest member of the group, Bryan was still not sure he belonged. He looked around at the others gathered at the giant "landing circle" on a hilltop near the compound. Like him, they were stamping their feet and breathing into their cupped hands to keep warm as they waited for the Leader to arrive with the important message they had been told to expect.

Soon the Leader strode up the hill and climbed a stepladder set up so that he could address the eighty or so faithful gathered to hear him. "The news we've been waiting for has come at last!" declared the Leader with a look of joy on his face. "The mother ship is returning to collect us! We must make ready. Salvation is from on high!"

The whole group cheered as the Leader kept on grinning and jabbing his finger at the sky. Bryan joined in the cheering and hugging. Yes, I do belong here, he was thinking happily. At last I have a family that loves me.

ONE THING becomes clear as one looks at the contactee phenomenon that emerged in the 1950s: it has strong connections to the occult. So perhaps it is not surprising that new religious movements, or cults, have developed out of the contactee phenomenon. Nor is it surprising that these UFO cults share the same occult influences as the contactee experience.

While the contactee movement reached its peak of popularity around 1960, a number of UFO-based religious groups coalesced around some of the contactees and their alleged messages from alien intelligences. Relatively small in terms of numbers of adherents, these UFO cults nevertheless remain popular in the United States and around the world.[1] Numbers may range from a few dozen members for a small group to possibly tens of thousands in some larger ones.[2] UFO cults typically fold or decline into obscurity when the founder dies, but some groups have endured from the 1950s. With names like "Brotherhood of the Seven Rays," "Earth Mission Planetary Outreach," and "Star Light Fellowship," a couple dozen identifiable UFO cults exist in America today, and there may be hundreds throughout the world.[3]

UFO cults may represent the strangest element in the world of UFO phenomena. Where did these groups come from? What do people who participate in them believe? What attracts them to such a cult? This chapter attempts to answer these questions by exploring the unusual world of UFO religion. Along the way, it provides a survey of four significant UFO cults and an evaluation of the worldview upon which such cults are based.

## Beliefs And Practices

EACH UFO CULT—like any other religious body—deserves to be considered individually. Each one is different in some respects from all the others. Yet it is helpful to consider some of the beliefs and practices that generally distinguish these UFO-related groups.

Most UFO cults share three interrelated beliefs: (1) Flying saucers are real. (2) People are in touch with alien intelligences associated with the flying saucers. (3) The messages given by aliens are of immense importance to human beings.

Persons connected with UFO cults may believe that flying saucers are physical spacecraft or they may believe that they have a nonphysical (psychic or spiritual) reality. Regardless, they believe UFOs to be a genuine reality. Indeed, the reality of extraterrestrial or interdimensional beings with their flying saucers forms the basis for the group's identity.

Members of such groups also believe that an individual or group of individuals is in touch with alien intelligences. This individual or group "channels" for these intelligences, that is, receives messages from them through paranormal means. Religion scholar and UFO researcher John A. Saliba explains:

> Many UFO groups have borrowed heavily from both spiritualism and Theosophy. They have incorporated in their ideology the concepts of cosmic wisdom and cosmic masters who exist on other planets. Their leaders often channel, or communicate with, these masters through some psychic means (such as telepathy) or by entering into a trance-like state.[4]

Saliba further identifies the practice of channeling alien intelligences or masters as the UFO cults' central feature.[5]

Cult members view these messages as prophetic and revelatory and consider them to be of immense technological, spiritual, and moral significance to the human race. Therefore, these messages are central to the belief of cult members. The role of the contactee, as well as the group as a whole, is to communicate and follow these revelatory messages. Typically, UFO cult followers believe that humanity's physical and spiritual welfare rests on whether or not they obey these revelations.

Beyond the three common and interrelated beliefs, other standard features can be found frequently in UFO cults.[6] Like many other religious cults, they typically have charismatic, if not eccentric, leaders at their center. Usually the central contactee, this leader offers an esoteric explanation for the UFO mystery. Religion researchers Robert S. Ellwood and Harry B. Partin describe what takes place in a typical UFO religion meeting:

> In some cases, individual contactees have delivered trance messages from UFOs in a manner virtually identical to the trance-preaching of Spiritualism. Another device has been the "circle," in which each individual in a small group adds something to a message the group believes it is collectively receiving from another world. Basically, however, UFOism is even more centered than Spiritualism upon the charismatic, shamanistic individual. Its

greatest events are focused upon them and are principally opportunities for these contactees to tell their story.[7]

Thus it's plain to see that mystical New Age beliefs and practices are common among UFO cults.

A venture into the stories of four of the best-known UFO cults reveals their bizarre nature. This journey also exposes the occult backgrounds of the leaders who established them.

## THE AETHERIUS SOCIETY

ONE OF THE oldest and best known of all UFO cults, the Aetherius Society has succeeded in attracting several thousand members from a number of countries. It exhibits many of the typical characteristics of UFO cults—founding by a contactee, obvious connections with occult beliefs and practices, and a mythology of alien species.

Former London taxicab driver George King (1919–1997) became the founder and leader of the Aetherius Society. Deeply involved as a young man in various metaphysical and occult groups and practices, including Theosophy, New Thought, Tantra Buddhism, Eastern mysticism (yoga), and mediumship,[8] King founded the society in 1954. Allegedly, an audible voice told him: "Prepare yourself. You are to become the voice of the Interplanetary Parliament." Then he claimed to begin receiving telepathic messages from a disembodied alien intelligence called "Master Aetherius." According to King, an interplanetary parliament of benevolent and concerned alien beings desired to communicate with the people of Earth and chose him as their "Primary Terrestrial Mental Channel."[9] King proclaimed himself the earthly ambassador to this alien parliament.

Regular messages supposedly came from the Venusian Master Aetherius as well as from several other cosmic masters within Earth's solar system, including "Master Jesus." The Aetherius Society views Jesus Christ as an advanced alien being from the planet Venus who communicates through a channel and travels to Earth in a flying saucer to protect Earth from evil forces. Ellwood and Partin further point out:

King began delivering, in trance, wisdom and instruction from Aetherius and other Masters such as Jesus and a Chinese saint, Goo-Ling. These Masters appear to be identical to the "Great White Lodge" of Theosophy, but saucers are regarded as the bearers of these beings or of their emissaries as well as of "magnetic" or spiritual power which is to be appropriated and radiated to the world.[10]

The benevolent goal of this "cosmic brotherhood" of ascended masters toward humankind is expressed through warnings concerning ecological and nuclear disasters and the provision of spiritual guidance. The aliens' greatest assistance to humanity may be their defense against nefarious alien forces. J. Gordon Melton explains:

> According to the Aetherius Society, earth is engaged in a cosmic warfare focused on the activities of certain "black magicians" seeking to enslave man. The cosmic brotherhood, the space hierarchy, wages war on these magicians. Members of the Society cooperate with the Brotherhood by channeling spiritual energy to particular concerns. Channeling activities are centered on certain periods when a space ship orbits earth and sends out special power.[11]

Most other UFO groups with origins in the 1950s have long since faded away. But not the Aetherius Society. This group's staying power may be due to the fact that it is well organized, with offices in Los Angeles and London. The organization also produces numerous publications—books, pamphlets, and periodicals—to keep its vision alive.

## THE UNARIUS ACADEMY OF SCIENCE

ONE OF THE most popular and outrageous UFO cults, Unarius was founded by Ernest L. and Ruth E. Norman in 1954. "Unarius" is an acronym for Universal Articulate Interdimensional Understanding of Science.

As in so many other cases, the founders of Unarius had roots in the occult. The Normans believed themselves to be the reincarnation of many great individuals of the past. Ernest, a spiritualist medium with prior involvement in flying saucer groups, identified himself as the reincarnation of Egyptian pharaoh Amenhotep IV, as Jesus Christ, and as a space traveler in ancient Atlantis.[12] Unarius followers described Ernest as "the greatest intelligence to ever come to earth." Not to be outdone by her husband, Ruth E. Norman revealed many remarkable past lives, claiming reincarnation as a pharaoh's mother, Confucius, Socrates, Mary Magdalene, Mona Lisa, Henry VIII, and others.[13]

Ernest declared an ability to channel and dictate messages from highly evolved alien beings living on other planets, especially Mars and Venus. Melton elaborates on Ernest's contactee mission: "Ernest's mission began when he materialized on earth and was guided by the evolved beings, now residing on other planets. From these teachers, he dictated seven books, which contain information about and teachings from the planets Venus, Mars, Hermes, Eros, Orion and Muse."[14]

When Ernest died (or as his followers say, his spirit moved to Mars) in 1971, Ruth continued advocating Eastern mystical and New Age beliefs and practices, channeling, reincarnation, and so on. She channeled otherworldly beings, among them "URIEL" (Universal Radiant Infinite Eternal Light). Later she insisted upon being called "Uriel: the Archangel and Cosmic Visionary." Besides channeling alien beings, Ruth also claimed to communicate with deceased scientists, statesmen, and various historical figures. The distinguished list of communicants includes Albert Einstein, Louis Pasteur, Dwight D. Eisenhower, John F. Kennedy, Herodotus, Aristotle, Mozart, Benjamin Franklin, Napoleon, and Henry Thoreau.

Ruth also claimed to be in contact with a confederation of alien beings from thirty-three planets. She predicted that a massive landing of spacecrafts would take place in the year 2001 on the Unarius property (sixty-seven acres) in El Cajon, California. This monumental landing would usher in the cosmic golden age of logic and reason.

In addition to her other exploits, Ruth conducted a healing ministry. Melton explains: "Healing is accomplished by Ray-Booms, the projected light beams from the great intelligences on the higher worlds."[15]

Likely the most eccentric of all UFO-based groups, the Unarius Academy of Science nevertheless continues today. Ruth E. Norman died at age ninety-two in 1996, but the organization carries on under the direction of Charles L. Spiegel.[16]

## HEAVEN'S GATE

THE UFO CULT came to the attention of most people only in 1997 when the news media reported that thirty-nine people had committed suicide in San Diego, believing their spirits would go to a UFO behind a comet. Heaven's Gate was a small cult, but it demonstrates the danger these new religions can pose.

At the center of the Heaven's Gate cult were two individuals: Marshall Herff Applewhite and Bonnie Lu Nettles.[17] Born in 1931, Applewhite was the son of a Presbyterian minister. He served as a professor of music at the University of St. Thomas and directed music at St. Mark's Episcopal Church, both located in Houston, Texas. Eventually, his admission of being sexually confused (having both heterosexual and homosexual relationships) led to scandal, loss of employment, and psychiatric problems.

Applewhite met Nettles in 1972 while he was hospitalized as a mental patient. Nettles worked at the hospital as a nurse. A member of the Theosophical Society, Nettles was deeply attached to occultism and practiced astrology, channeling, and interaction with spirit guides. Immediately upon meeting Applewhite in the hospital, she developed a platonic relationship with him, and their deep spiritual bond lasted until Nettles's death in 1985.

In 1973, Applewhite and Nettles began traveling through western states, calling themselves "Bo" and "Peep" (later "Do" and "Ti," after the musical tones) and teaching their brand of UFO religion. They believed and taught that they were incarnate extraterrestrials who channeled important information to human beings. Sociologist Robert W. Balch describes the basic worldview held by most members:

Most of the people who joined the UFO cult were spiritual seekers who shared a worldview where reincarnation, lost continents, flying saucers, and psychic phenomena were taken for granted. Today this worldview is called New Age, but in the 1970s, before that term became commonplace, people were more likely to say they were "into metaphysics."[18]

The movement began to grow.

Applewhite and Nettles recruited followers by convincing them that they were the two witnesses written about in the book of Revelation in the Bible. They claimed they would fulfill Revelation 11:3–12 by being killed, lying dead in the street for three days, and then being raised from the dead into a spaceship, along with all of their followers. They predicted a public appearance of flying saucers in the mid-1970s, later claiming that the spaceships had in fact appeared but were visible only to the true believers. The two later also explained to their followers a revelation that they were no longer required to die and rise again.

The group—at one point called Human Individual Metamorphosis (HIM)[19]—was comprised of between 150 and 200 members at its peak. Expected to give up everything (family, friends, and possessions), the followers were devoted completely to the teachings of "the two": Applewhite and Nettles. Sexual expression among members was strongly discouraged, and Applewhite himself underwent castration surgery to relieve his sexual tension. Balch, who infiltrated the group, explains its UFO-oriented message:

> They explained that "the Father's kingdom" is not a spiritual realm as Christians believe, but the "literal heavens," meaning the entire universe, and the only way to get there is in a spacecraft. Earth is just one of countless heavenly "gardens" that members of the next level planted with the seeds of consciousness millions of years ago. Now, they said, this garden is being prepared for its first and possibly last "harvest." Only by escaping the planet's spiritually poisoned atmosphere could humans expect to break

the endless cycle of death and reincarnation. Once seekers reached the next level, they would become immortal, androgynous beings living in a state of perpetual growth.[20]

By the late 1970s, the group became more secretive and dropped out of sight. Nettles died of cancer in 1985, leaving Applewhite to lead the group alone. In 1993 the group took out ads and called itself Total Overcomer's Anonymous.[21] Then, in 1995, the group changed its name to Heaven's Gate and moved to the San Diego, California area. The group started a successful business designing commercial Web sites. In October 1996 the group moved into a mansion in an upscale San Diego neighborhood.

Applewhite and his converts believed that a spacecraft was following the Hale-Bopp comet. In videotapes the group explained that they were putting off their earthly vehicles by committing suicide in order to join this flying saucer and take the next step in human evolution.[22] On March 26, 1997, the members of the Heaven's Gate UFO cult committed suicide en masse. Thirty-nine androgynous-looking individuals (eighteen men and twenty-one women), dressed exactly alike, took a lethal combination of barbiturates and vodka and pulled plastic bags over their heads.[23] Applewhite and his followers committed suicide based upon their faith in their UFO religion. They were willing to make the ultimate sacrifice for their belief system.

## THE RAELIAN MOVEMENT

THE RAELIAN MOVEMENT—so-called because of the adopted name of its founder, Rael—is well known not only because it is apparently the largest UFO cult but also because of its reputation for sexual promiscuity. This group shows what effect UFO beliefs can have on morals and behavior.

This movement came into being because of one man: Claude Vorilhon (b. 1946).[24] A former race car driver and journalist, Vorilhon claims to have encountered space aliens on December 13, 1973, while walking near a volcanic range in Clermont-Ferrand, France. In his booklet *Help Us Welcome Extraterrestrials*, he describes the aliens as small, humanoid beings with pale green skin and almond eyes. These aliens, known as the "Elohim" (which also happens to be a Hebrew word for "God"), allegedly transported Vorilhon to their planet, where he observed the Elohim civilization firsthand and even engaged in sexual relations with several female humanoid robots.[25]

In his book *Let's Welcome Our Fathers from Space: They Created Humanity in Their Laboratories,* Vorilhon explains that the Elohim are superior extraterrestrial scientists who created humankind in their image inside their DNA laboratory and subsequently "implanted" them on Earth.[26] The Elohim chose Vorilhon, whom they called Rael, and entrusted him with their revealed messages as their only authentic contact. In typical contactee fashion, Vorilhon claimed the Elohim were speaking directly through him (Rael) and they revealed him to be half alien (the seed of Elohim and a mortal woman) and "the last of forty prophets."[27] The other thirty-nine immortal prophets include such distinguished individuals as Jesus, Buddha, Muhammad, Joseph Smith, and others, who are also half alien.[28]

In accord with his cosmic calling, Rael founded the Raelian movement in 1973. Carrying out the mission of the Elohim, the movement is to warn humanity of the dangers of a possible nuclear holocaust ("Age of Apocalypse") and promote an advanced planetary consciousness ("Age of Revelation"). The movement plans to construct a Raelian embassy in Jerusalem by the year 2025 in order to usher in the coming of the Elohim, who will descend to Earth aboard flying saucers.[29]

While rejecting traditional theistic concepts of God, the Raelians do embrace an alleged scientific view of immortality. They believe that the Elohim will allow them to achieve immortality through the cloning of individual followers. Raelian initiates must even sign a contract that allows a part of their forehead bones (the "third eye") to be removed and stored in ice after their deaths as they await the return of the Elohim.[30]

The Raelian movement has drawn considerable media attention for its permissive views concerning sexual expression. While rejecting a traditional approach to marriage and family because of population overgrowth, the Raelians encourage sexual ambiguity, homosexuality, and unrestrained sexual pursuits.[31] Rael, often photographed in the company of scantily dressed women, is suspected of living a playboy's lifestyle. He openly promotes or condones homosexuality, abortion, premarital sex, out-of-wedlock childbirth, and virtually all sexual exploration.[32] The Raelian movement appeals to the so-called sexually liberated. With an estimated membership of from twenty to thirty thousand, it has offices in France, Japan, Canada, Mexico, Africa, and the United States.[33]

Obviously, the beliefs of UFO cults such as the Raelian movement extend well beyond the normal bounds of intellectual credibility. Yet these groups show no sign of going away. Their unique mix of scientized myth and occult religion appeals to many people. UFO cults, despite their bizarre beliefs and practices, meet some of the needs of spiritually starving people.

## THE UFO WORLDVIEW

IN EVALUATING UFO cults, as well as the sort of abductee and contactee experiences on which they are based, what stands out is a widespread oblivion, on the part of UFO believers, to the importance of an adequate, carefully considered, and solidly constructed worldview. According to philosopher Ronald H. Nash, a worldview is "a conceptual scheme" for interpreting reality.[34] The attempt to arrange one's most basic beliefs into a coherent system enables an individual to evaluate and interpret information and experience. A carefully considered worldview protects a person from irrational, even dangerous, thoughts and actions.

A belief that space aliens literally and physically exist comports well with a belief system one may call the UFO-extraterrestrial (UFOET) worldview. How do the abductee and contactee experiences and UFO religions look after one has examined the worldview on which they depend?

When considering any worldview, the following tests offer invaluable tools for evaluation.[35]

- *Coherence test.* Is a particular worldview logically consistent? An acceptable worldview avoids "self-stultification," or incon-

sistencies, and includes component parts that hang together as a coherent whole.
- *Mean test.* Is the worldview balanced between complexity and simplicity? An acceptable worldview will be neither too simple (the reductive fallacy) nor too complex (Ockham's razor). All things being equal, the simplest, most economical, and yet fully orbed worldview serves a person best.
- *Explanatory power and scope test.* How effectively and comprehensively does a worldview account for reality? An acceptable worldview accounts for both a depth and a breadth of questions about reality.
- *Correspondence test.* Does a particular worldview correspond with well-established, empirical facts? An acceptable worldview matches the facts of the observable world.
- *Verification test.* Can the central truth claims of the worldview be verified or falsified? An acceptable worldview relies on claims that can be tested and proven true or false.
- *Pragmatic test.* Does the worldview yield practical applications? An acceptable worldview will be workable, sensible, and therefore "externally livable."
- *Existential test.* Does the worldview address the internal needs of humanity? An acceptable worldview accounts for humans' need for meaning, purpose, and significance and therefore is "internally livable."
- *Competition test.* Can a worldview successfully compete in the marketplace of ideas? An acceptable worldview responds to reasonable challenges and offers a reasonable critique of competing worldviews.
- *Predictive test.* Can a worldview successfully anticipate future discoveries? An acceptable worldview can help humans make some accurate predictions of what their research into the macrocosm and microcosm will yield. It accommodates and incorporates emerging data.

The weaknesses of the UFO-ET worldview stand out. Such a view proves a woefully inadequate basis for judging reality. Many people fail to recognize and respond to this inadequacy, and cultists prey upon peoples' vulnerability to intellectual, emotional, and spiritual deception. It is important, consequently, that people step back and take a look at UFO phenomena and extraterrestrial intelligence from a biblical viewpoint.

CHAPTER 15

# THE BIBLE AND UFOS

*Hugh Ross and Kenneth Samples*

I saw a wheel on the ground beside each creature.... This was the appearance and structure of the wheels: They sparkled like chrysolite, and all four looked alike. Each appeared to be made like a wheel intersecting a wheel. As they moved, they would go in any one of the four directions the creatures faced; the wheels did not turn about as the creatures went. Their rims were high and awesome, and all four rims were full of eyes all around.

When the living creatures moved, the wheels beside them moved; and when the living creatures rose from the ground, the wheels also rose. Wherever the spirit would go, they would go, and the wheels would rise along with them, because the spirit of the living creatures was in the wheels. (Ezekiel 1:15-20)

SINCE THE dawn of the flying saucer era, people have argued that the Bible describes UFO phenomena. In addition to Ezekiel's astonishing vision of wheels within wheels (Ezekiel 1:4-28; 10), the Bible tells of a fiery chariot and horses whisking Elijah away from Earth (2 Kings 2:1-12). Daniel and John both experienced amazing visions of powerful beings and magnificent places (Daniel 7:2-12; 8:1-14; Revelation 1:12-18). The Magi followed an unusual "star" all the way from the East to Jerusalem and then to the very house where the young Jesus lived (Matthew 2:1-12).

Given that the extradimensional hypothesis represents a scientifically and biblically credible view of reality, some connection can be seen between these scriptural accounts and residual UFOs (RUFOs). Angels—that is, ministering spirits who have not rebelled but faithfully serve God—may become visible to humans in certain situations. Like some RUFOs, they are real but not physical. In the case of the Bethlehem star, the object may actually have been an IFO, such as a recurring nova or some other natural phenomenon employed by God with perfect timing and placement.[1] But each of the other phenomena fits the profile of an emissary to space-time from beyond space-time. Each encounter turns worship and attention toward God rather than away from Him. None yields a message that contradicts any of the rest of Scripture. Thus, they starkly contrast with the RUFOs of abductee, contactee, and cult phenomena. If the extra dimensional hypothesis is correct, then it would appear that the similarity between angel appearances in the Bible and RUFOs is owing to the fact that the RUFOs represent the angels' wicked peers: the demons. Demons are spirit beings who presumably can operate beyond these four dimensions but have very different intentions for humans than do the good angels.

But what if one prefers the extraterrestrial hypothesis over the extra dimensional hypothesis? Does the Bible support the idea that there may be physical life on other planets? This chapter discusses that question as well as how non-Christians see the connection between UFOs and God. People can see what dangers the Bible identifies in occult behavior and how they can protect themselves from the evil influence of UFOs through putting on the "armor of God." Finally, a principle will be considered—that all people who respect the scientific method, whether Christian or not, should be able to agree upon and use.

## Extraterrestrial Life

THOSE WHO ACCEPT the authority of the Bible and embrace a Christian worldview take different positions on whether God might have created intelligent life on other planets. This question has been debated at least since Thomas Aquinas discussed it nine centuries ago.

Scholars who believe extraterrestrial intelligence (ETI) physically exists see it as a display of God's creativity and power. They argue that a God who so obviously enjoys creating, a God of unimaginable power, should not be expected to limit His creative expression to just one planet and its one species of spiritual beings.

This is defensible reasoning. Nevertheless, both physical and nonphysical data argue against such an extraterrestrial hypothesis for UFOs. As learned in chapter 5, the laws of physics make travel by intelligent aliens to this solar system virtually impossible.

Furthermore, from the spiritual side, one might ask what purpose would be served by E.T.'s traveling to Earth if E.T. did exist. The Creator has already provided humans with everything needed, materially and spiritually, for life on this planet—and for immortality.

He has promised that partial knowledge, though adequate for current circumstances, will be traded for more complete knowledge when His plans for this world have been fulfilled. Until then, wisdom and counsel are available for the asking (James 1:5).

Whether or not God has chosen to express His creativity by placing life—any kind of physical life—in various locations of the cosmos, God will create again, according to the Bible, when His conquest of evil is complete. The book of Revelation, as well as other parts of the Bible, forecast in some detail the "new creation" (Revelation 21–22). The closing words of Scripture ring with that promise.

## SECULARISTS' VIEWS

C. S. LEWIS once dryly commented on atheists' attempts to use both sides of the ETI question to argue against the Christian worldview:

> If we discover other bodies, they must be habitable or uninhabitable: and the odd thing is that both these hypotheses are used as grounds for rejecting Christianity. If the universe is teeming with life, this, we are told, reduces to absurdity the Christian claim—or what is thought to be the Christian claim—that man is unique, and the Christian doctrine that to this one planet God came down and was incarnate for us men and our salvation. If, on the other hand, the earth is really unique, then that proves that life is only an accidental by-product in the universe,

and so again disproves our religion. Really, we are hard to please.[2]

Deists and agnostics seem deeply troubled by the apparent loneliness of humanity in the vast cosmos of ten billion trillion stars. Britain's famed physicist Stephen Hawking finds it "very hard to believe" that God would make so many "useless" stars if His intent were to make a home for humanity.[3] Carl Sagan repeated this theme in his lectures and books, and it was popularized in the movie *Contact:* if humans are the only sentient, civilized beings in the universe, the universe is all just a waste of space.

Because the biblical God must abhor waste, say Hawking and Sagan, one of two conclusions must follow: either He is not the Creator or, if He is, the universe must be teeming with life. An accumulation of research findings suggests a third alternative: the biblical God invested the entirety of the cosmos in creating one planet in one place at one time suitable for life, then placed life upon it, all to accomplish a purpose so awesomely wonderful that humans can barely begin to comprehend it. As chapter 9 tells, given the physical laws of the universe, 10 billion trillion stars, no more, no less, are essential for the existence of one planet in one era capable of supporting physical life.

Because the biblical God desires to enhance love and minimize suffering, the question must be asked: How could physical life-forms from other planets help accomplish such a goal? As created beings, their knowledge, power, and love could never equal those of the Creator Himself.

For these reasons and many others presented in the preceding chapters, the authors of this book believe that the extraterrestrial hypothesis does not stand up. UFOs and extraterrestrials (when they are not hoaxes or misperceptions) must instead be supernatural creatures of a malevolent nature.

## SPIRITUAL DANGERS

PARTICIPATION WITH UFO phenomena opens the door, whether a

person recognizes it or not, to occult phenomena. According to historical Christian theology, such phenomena carry negative consequences for a person's overall health and vitality in this life and the next.

Scripture expressly forbids involvement in any occult practices. "Let no one be found among you ... who practices divination or sorcery, interprets omens, engages in witchcraft, or casts spells, or who is a medium or spiritist or who consults the dead. Anyone who does these things is detestable to the LORD" (Deuteronomy 18:10-12). UFO experiences involve such occult beliefs and practices as mediumistic "channeling" of alien (psychic or spirit) entities, automatic writing, telepathy, teleportation, dematerialization, levitation, and psychokinesis.[4]

To engage in this hidden world is to seek knowledge—and power—apart from God's intended and freely given revelation of truth. Thus, occult practitioners submit themselves to spiritual authority other than God's, with mentally, emotionally, and spiritually devastating results.[5]

The area of UFO phenomena that perhaps most clearly reflects the occult is the trance channeling performed by contactees. The Bible warns, "Do not turn to mediums or seek out spiritists, for you will be defiled by them" (Leviticus 19:31). Deeply rooted in pagan religion, trance channeling may be described as spiritism with a new twist.

Theologian and apologist Ron Rhodes defines a channeler as "a person who yields control of his or her perceptual and cognitive capacities to a spiritual entity or force with the intent of receiving paranormal information."[6] Thus, contactees are seeking revelation from a divinely forbidden source (Exodus 22:18; 2 Kings 21:6; Ezekiel 13:9; Zechariah 10:2). If and when real contact is made, that link connects the person with what the Bible calls "the powers of this dark world" (Ephesians 6:12). Because channelers allow entities to take over their minds and bodies, they are susceptible to, and may indeed become, possessed by demons—a dreadful risk (Matthew 4:24; 9:32-34; 12:22-24; Mark 3:10-12; Luke 11:14-26).

This emphasis on darkness, danger, and malevolence does not mean that abductee or contactee experiences necessarily appear horrific. Some appear, at least initially, as beautiful and uplifting encounters. People often wonder how such a positive experience or message can come from any source other than a good and godly one. The Bible anticipates this question. The apostle Paul warned his friends that a messenger of darkness often "masquerades as an angel of light" (2 Corinthians 11:14-15). Such beings can perform convincing and deceptive "miracles, signs and wonders" (2 Thessalonians 2:9-10; see also Matthew 24:24).

Some contactees declare Jesus Christ to be an extraterrestrial and claim to receive channeled messages from Him.[7] This claim represents a direct challenge to the historic Christian claim that God has already revealed adequate truth in Scripture (1 Corinthians 4:6; 2 Timothy 3:16; 2 Peter 1:20-21). The foundational truths of Christianity have been "once for all entrusted to the saints" (Jude 3). No need exists for further revelation, and no revelation from God or His messengers will contradict the revelation of Scripture (2 Corinthians 11:4; Galatians 1:8-9; 1 Timothy 4:1; 6:3; Titus 1:9; 2 Peter 2:1; 1 John 2:22-23; 2 John 7-11).

Furthermore, the metaphysical messages and revelations upon which virtually all UFO cults are based directly deny and contradict historic Christian doctrines about God, Christ, sin, salvation, and Scripture. UFO religions, such as those based on the revelations of *The Urantia Book*, categorically reject orthodox Christology (Jesus' identity as both God and man) and thus reject Jesus Christ as the cosmic Creator and resurrected Savior of humankind.

The theological messages promulgated by the Urantia organization and the Aetherius Society tend to be pluralistic (all religions are true), monistic (all reality is one), universalistic (no divine judgment is coming), and mystical, thus matching more closely with New Age mysticism than with the teaching of Scripture.[8] Members of UFO cults typically affirm distinctive New Age beliefs and practices, including reincarnation, channeling, telepathy, and human evolution toward godhood.

The worldview expressed in UFO-based religion leaves no room for such biblical distinctives as humankind's creation in the image of God, original sin and humankind's resultant need for repentance, and humankind's absolute dependency on the grace of God. As a whole, the religious worldview characteristic of UFO cults differs fundamentally from the worldview of historic Christian theism.

## DEFENSES AGAINST UFOS

IT'S TRUE THAT demons are powerful and dangerous. But people need not feel hopeless in the face of their attacks. There is a greater power at hand: God's. And by His gracious provision, God makes His power available to human beings through the agency of His Son, Jesus Christ.

Any defense against residual UFOs must be broad enough to deal with the whole spectrum of attacks that demons mount through the medium of UFO encounters. Sometimes the RUFOs may be the demons themselves. Sometimes demons may so disturb the spiritual and psychological state of some humans that they experience visions or hallucinations interpreted as UFOs. Demons also have the power to indirectly manifest RUFOs through possessing humans or in some way oppressing them.

Obviously, demon oppression can be manifested in people to different degrees. Moreover, a fine line separates demon oppression from the tendency in all people to sin and rebel against the authority of God. However, Christian workers have observed that it takes a significant level of demon oppression for close encounters with residual UFOs to occur.[9]

According to the Bible, God has provided all the defenses necessary to prevent demonic deception and attacks. In other words, no one need ever suffer from demonic oppression. But the Bible also acknowledges that humans are notoriously derelict in putting on "the full armor of God" (Ephesians 6:11). Often people either close their eyes and minds to the dangers or else think they are strong enough to face such perils on their own. Others believe demonic visitors who say they only want to help humans gain more knowledge and power.

The full armor of God, as described and explained in the sixth chapter of Paul's letter to the Ephesians, not only can protect believers in Jesus Christ against the attacks and deceptions of demons, but also can equip believers for helping others escape spiritual deception and demonic depression.

> Be strong in the Lord and in his mighty power. Put on the full armor of God so that you can take your stand against

the devil's schemes. For our struggle is not against flesh and blood, but against the rulers, against the authorities, against the powers of this dark world and against the spiritual forces of evil in the heavenly realms. Therefore put on the full armor of God, so that when the day of evil comes, you may be able to stand your ground, and after you have done everything, to stand. Stand firm then, with the belt of truth buckled around your waist, with the breastplate of righteousness in place, and with your feet fitted with the readiness that comes from the gospel of peace. In addition to all this, take up the shield of faith, with which you can extinguish all the flaming arrows of the evil one. Take the helmet of salvation and the sword of the Spirit, which is the word of God. And pray in the Spirit on all occasions with all kinds of prayers and requests. With this in mind, be alert and always keep on praying for all the saints. (Ephesians 6:10-18)

The defenses God has provided against demonic oppression and deception are neither complex nor difficult to implement. If you have been troubled by UFO phenomena, here are some steps you can take: (1) Examine your life for any open invitations to demonic oppression and deception. (2) Repent of making those invitations. (Repenting means agreeing with God that providing invitations to demons was wrong and seeking His help to permanently turn away from such invitations.) (3) Confess to God that the invitations to demons provided by your parents or other close relatives were wrong. (4) Take action to demonstrate repentance and to purify your life, including repudiating and renouncing all occult involvement. (5) Turn your life over to Jesus Christ, accepting His offer to forgive all of your sins and giving Him the complete authority He deserves. (6) Acknowledge privately and publicly that His will, not your own, is supreme. (7) Dedicate your heart, soul, mind, and strength to be used for the fulfillment of His purposes.

## A Point Of Agreement

THE AUTHORS OF THIS book are fully convinced that both science and Scripture point to the supradimensional beings known as demons as the malevolent sources of RUFO phenomena. At the same time, these writers know some others will continue to believe the extraterrestrial hypothesis or other explanations. Yet amid a host of diverse and often confusing facts and opinions about UFO phenomena, at least one point of agreement stands out among physical scientists, social scientists, behavioral scientists, and students of the Bible: *pursue truth where it may be found and tested, and beware of any pursuit that links, directly or indirectly, with occult power or "secret" knowledge.* History, both ancient and recent, both Eastern and Western, provides ample evidence in support of a great irony: pursuit of power leaves a person vulnerable to deception and despair, while pursuit of truth leads to strength and hope—and an abundant life (see John 10:10).

The motto inscribed in stone at the west entrance to the Caltech campus reads, "The truth will set you free." These words, taken from the Bible (John 8:32), apply to every human being. They provide enough challenge, enough power, and enough protection for any human lifetime.

CHAPTER 16

# SUMMARY

*Hugh Ross*

LIGHTS IN the sky and little green men—speculations about them may always persist. But solid answers to questions about them do exist. Considering the issues in relationship to physical science, theology, philosophy, psychology, sociology, and political science can help solve the mysteries.

Ever since the beginning of the "flying saucer age" in 1947, and even well before that, people have observed what appear to be mysterious aerial visitors. The nature of these phenomena makes them difficult, but not impossible, to study. Investigators have developed classification systems to help them in sifting through the multitude of UFO reports. Researchers are able to offer standard explanations for most of the sightings—the UFOs are hoaxes or fireballs or the planet Venus or the lights of an aircraft, for example. But a small percentage of UFOs remains unexplained after all the natural explanations have been exhausted. These are the residual UFOs, or RUFOs. Ufologists usually resort to one of two hypotheses to explain RUFOs: the extraterrestrial hypothesis (RUFOs are physical visitors from outer space) and the interdimensional hypothesis (RUFOs are entities from beyond physical dimensions).

If an inquirer looks into the extraterrestrial hypothesis, or ETH, from a scientific standpoint, that person quickly realizes that it does not stand up.

Where would extraterrestrial intelligence come from? There are fewer planets in the universe than many people presume. Furthermore, the vast and growing number of known characteristics required for a planet to support life essentially rules out the possibility that a suitable home for physical life can be found anywhere in the universe but here on Earth.

Nor is it possible that life could have arisen on other planets according to the principles of evolutionary theory. The likelihood that chemicals anywhere in the universe would assemble into a living organism is nil.

Apart from these facts, even if E.T. somehow did exist somewhere, traveling to Earth would pose insurmountable obstacles for him (or it). Vast distances necessitate navigating at high speeds through obstacles such as space dust, meteors, comets, radiation, and gravitational perturbations. Thousands of years of high-velocity travel pose energy and shielding problems that are impossible to resolve. And wormholes do not provide shortcuts because anyone passing through one would be obliterated!

Yet RUFOs do exist. They are clearly real, though they do not obey the laws of physics. Could it be that they would be better understood if people had access to all the information known by the government? Most unlikely. The government would be unable to contain such explosive information as evidence for the existence of aliens. The appearance of cover-up can be explained by normal bureaucratic behavior. The idea of a government conspiracy perpetuates itself when people develop a theory and then make all the evidence fit their theory.

In the end, it seems best to abandon the ETH and consider the second option: the interdimensional hypothesis, or IDH. While in its more popular versions the IDH departs from verified and verifiable reality, the extradimensional hypothesis holds promise. And here the investigator stands on firmer ground, scientifically speaking. The space-time theorems show that supernature must exist, because nature was plainly originated by something beyond itself. Therefore, a scientifically credible possibility exists that RUFOs come from beyond the four familiar dimensions of the universe.

If one takes the extradimensional hypothesis to mean that entities could come into the universe from a spiritual realm, one can see a remarkable correspondence between science and Scripture. The Bible describes a Creator who is beyond matter, energy, and the space-time dimensions of the universe. It also describes spirit beings who are able to enter the universe and exhibit physical effects.

A closer examination of RUFOs shows that they are consistent with the Bible's descriptions of demons. The RUFOs appear to be alive and to be acting in an intelligent way with malevolent intentions.

While no published studies have formally tested for this demonic connection, considerable anecdotal evidence suggests that people who encounter these residual phenomena have previously opened themselves to the forces of evil. How? By participating, knowingly or unknowingly, in occultism or occult-related activities—for example, fortune telling, Ouija board games, séances, and so on—all of which are biblically forbidden.

The connection with the occult is plain to see in the case of those who have the closest encounters with UFOs: abductees and contactees. Abductees are people who tell a story of being captured by aliens and taken aboard spacecraft for examination. Contactees are people who claim to be used by aliens as channels of information. These reported experiences are completely in line with such occult practices as trance channeling of "ascended masters."

From out of contactee experiences have arisen several UFO cults, which are best understood as new religious movements of an occult nature. These cults claim to possess information from aliens that humanity needs to know. In addition to their bizarre beliefs, these cults often engage in immorality—and may even become deadly, as in the case of the Heaven's Gate cult. The worldview held by cult members and other UFO believers isn't logical or coherent, but the participants have such a powerful spiritual need that they are willing to accept it anyway.

The biblical worldview is much more satisfying. It leaves open the possibility that God could have created life elsewhere in the universe, but it also suggests that there is no reason for Him to have done so. The Bible warns of the danger of occult practices, such as those associated with UFOs, and from the Bible one learns how to find protection from the evil spirits who are causing RUFO encounters. The Bible encourages a pursuit of the truth in this area as in all others.

The truth about UFOs can be known. Indeed, the UFO mystery is a mystery solved. Earth is not being visited by aliens from another planet, but some people are being visited by spirit beings who want everyone to *think* they are aliens from another planet. By trusting the revelation given by the greatest transdimensional Being of them all, people need never wonder about UFOs again. When people put their lives in the hands of this Cause of human existence, this God who loves every person, the fear of UFO demons and what they can do evaporates.

APPENDIX A

# Fine-Tuning for Life on Earth

REMARKABLE FINE-tuning must take place in order for life to exist on any planet in any planetary system. Earth exists in the only known planetary system able to demonstrate all the characteristics required for life to survive. The following essential characteristics for life show Earth's matchless fine-tuning within its parameters.[1]

1. Galaxy cluster type
    - If Earth's galaxy cluster were too rich, galaxy collisions and mergers would disrupt the solar orbit.
    - If Earth's galaxy cluster were too sparse, there would be insufficient infusion of gas into the Milky Way to sustain star formation there for a long enough period of time.

2. Galaxy size
    - If the Milky Way were too large, infusion of gas and stars would disturb the sun's orbit and ignite too many galactic eruptions.
    - If the Milky Way were too small, there would be insufficient infusion of gas to sustain star formation for a long enough period of time.

3. Galaxy type
    - If the Milky Way were too elliptical, star formation would have ceased before sufficient heavy elements had built up for life chemistry.
    - If the Milky Way were too irregular, radiation exposure on occasion would be too severe and heavy elements for life chemistry would not be available.

4. Galaxy mass distribution
    - If too much of the Milky Way's mass resided in the central bulge, the Earth would be exposed to too much radiation.
    - If too much of the Milky Way's mass resided in the spiral arms, the Earth would be destabilized by the gravity and by radiation from adjacent spiral arms.

5. Galaxy location
    - If the Milky Way were located too close to a rich galaxy cluster, the Earth would be gravitationally disrupted.
    - If the Milky Way were located too close to a very large galaxy (or galaxies), the Earth would be gravitationally disrupted.

6. Supernovae eruptions
    - If supernovae had occurred too close, life on Earth would be exterminated by radiation.
    - If supernovae had occurred too far away, there would not be enough heavy element ashes for the formation of rocky planets like Earth.
    - If supernovae had occurred too infrequently, there would not be enough heavy element ashes for the formation of rocky planets.
    - If supernovae had occurred too frequently, life on Earth would be exterminated.
    - If supernovae had occurred too soon, there would not have been enough heavy element ashes for the formation of rocky planets.
    - If supernovae had occurred too late, life on Earth would be exterminated by radiation.

7. White dwarf binaries
    - If there were too few white dwarf binaries, there would be insufficient fluorine for life chemistry.
    - If there were too many white dwarf binaries, planetary orbits would be disrupted by stellar density and life on Earth would be exterminated.
    - If white dwarf binaries had appeared too soon, there would not

be enough heavy elements for efficient fluorine production.
- If white dwarf binaries had appeared too late, fluorine would be made too late for incorporation in Earth's protoplanet.

8. Proximity of solar nebula to a supernova eruption
   - If the solar nebula were farther away, the Earth would have absorbed insufficient heavy elements for life.
   - If the solar nebula were closer, the nebula would be blown apart.

9. Timing of solar nebula formation relative to supernova eruption
   - If the solar nebula had formed earlier, the nebula would have been blown apart.
   - If the solar nebula had formed later, the nebula would not have absorbed enough heavy elements.

10. Number of stars in parent star birth aggregate
    - If there were too few stars in the parent star birth aggregate, there would have been insufficient input of certain heavy elements into the solar nebula.
    - If there were too many stars in the parent star birth aggregate, planetary orbits would be too radically disturbed.

11. Star formation history in parent star vicinity
    - If there had been too much star formation going on in the vicinity of the sun, planetary orbits would be too radically disturbed.

12. Birth date of the star-planetary system
    - If the system had been born too early, the quantity of heavy elements would have been too low for large, rocky planets like Earth to form.
    - If the system had been born too late, the sun would not yet have reached its stable burning phase. Furthermore, the ratio of potassium-40, uranium-235, uranium-238, and thorium-232 to iron would be too low for long-lived plate tectonics to be sustained on Earth.

13. Parent star distance from center of galaxy
    - If the sun were too far from the center of the galaxy, the quantity of heavy elements would have been insufficient to make rocky planets like Earth. In addition, there would be the wrong abundances of silicon, sulfur, and magnesium relative to iron for appropriate planet core characteristics.
    - If the sun were too close to the center of the galaxy, galactic radiation would be too great and stellar density would disturb planetary orbits. Again, there would be the wrong abundances of silicon, sulfur, and magnesium relative to iron for appropriate planet core characteristics.

14. Parent star distance from closest spiral arm
    - If the distance were too great, the quantity of heavy elements would be too small for rocky planets to form.
    - If the distance were too small, the solar system would experience gravitational disturbances and radiation exposure.

15. Z-axis heights of star's orbit
    - If the z-axis height were too high, exposure to harmful radiation from the galactic core would be too great.

16. Number of stars in the planetary system
    - If there were multiple stars in the solar system, tidal interactions would disrupt Earth's orbit.
    - If there were no stars in the system, Earth would have insufficient heat to support life.

17. Parent star age
    - If the sun were older, its luminosity would change too quickly.
    - If the sun were younger, its luminosity would change too quickly.

18. Parent star mass
    - If the sun's mass were greater, its luminosity would change too quickly and it would burn too rapidly.
    - If the sun's mass were smaller, the range of planet distances

that would make life possible would be too narrow. In addition, tidal forces would disrupt Earth's rotational period. Also, ultraviolet radiation would be inadequate for plants to make sugars and oxygen.

19. Parent star metallicity
    - If the sun's metallicity were too small, there would be insufficient heavy elements for life chemistry.
    - If the sun's metallicity were too large, life would be poisoned by heavy-element concentrations. Furthermore, radioactivity would be too intense for life.

20. Parent star color
    - If the sun were redder, photosynthetic response would be insufficient.
    - If the sun were bluer, photosynthetic response would be insufficient.

21. Galactic tides
    - If galactic tides were too weak, the comet ejection rate from the giant planet region would be too low.
    - If galactic tides were too strong, the comet ejection rate from the giant planet region would be too high.

22. $H_3^+$ production
    - If $H_3^+$ production had been too small, simple molecules essential to planet formation and life chemistry would not form.
    - If $H_3^+$ production had been too large, planets would form at the wrong time and place for life.

23. Flux of cosmic ray protons
    - If the proton flux had been too small, there would be inadequate cloud formation in Earth's troposphere.
    - If the proton flux had been too large, there would be too much cloud formation in Earth's troposphere.

24. Solar wind
    - If the solar wind were too weak, too many cosmic ray protons would reach Earth's troposphere, causing too much cloud formation.
    - If the solar wind were too strong, too few cosmic ray protons would reach Earth's troposphere, causing too little cloud formation.

25. Parent star luminosity relative to speciation
    - If the sun's luminosity had increased too soon, a runaway greenhouse effect would develop on Earth.
    - If the sun's luminosity had increased too late, runaway glaciation would develop on Earth.

26. Surface gravity (escape velocity)
    - If surface gravity were stronger, Earth's atmosphere would retain too much ammonia and methane.
    - If surface gravity were weaker, Earth's atmosphere would lose too much water.

27. Distance from parent star
    - If Earth's distance from the sun were greater, Earth would be too cool for a stable water cycle.
    - If Earth's distance from the sun were lesser, Earth would be too warm for a stable water cycle.

28. Inclination of orbit
    - If Earth's orbital inclination were too great, temperature differences would be too extreme.

29. Orbital eccentricity
    - If Earth's orbital eccentricity were too great, seasonal temperature differences would be too extreme.

30. Axial tilt
    - If Earth's axial tilt were greater, surface temperature differences would be too great.

- If Earth's axial tilt were lesser, surface temperature differences would be too great.

31. Rate of change of axial tilt
    - If Earth's rate of change of axial tilt were greater, climatic changes and surface temperature differences would be too extreme.

32. Rotation period
    - If Earth's rotation period were longer, diurnal temperature differences would be too great.
    - If Earth's rotation period were briefer, atmospheric wind velocities would be too great.

33. Rate of change in rotation period
    - If the rate of change in Earth's rotation period were more rapid, the surface temperature range necessary for life would not be sustained.
    - If the rate of change in Earth's rotation period were less rapid, the surface temperature range necessary for life would not be sustained.

34. Planet age
    - If the Earth were too young, it would rotate too rapidly.
    - If the Earth were too old, it would rotate too slowly.

35. Magnetic field
    - If the Earth's magnetic field were stronger, electromagnetic storms would be too severe. Also, too few cosmic ray protons would reach the Earth's troposphere, and this would inhibit adequate cloud formation.
    - If the Earth's magnetic field were too weak, the ozone shield would be inadequately protected from hard stellar and solar radiation.

36. Thickness of crust
    - If the Earth's crust were thicker, too much oxygen would be transferred from the atmosphere to the crust.

- If the Earth's crust were thinner, volcanic and tectonic activity would be too great.

37. Albedo (ratio of reflected light to total amount falling on surface)
    - If the Earth's albedo were greater, runaway glaciation would develop.
    - If the Earth's albedo were smaller, a runaway greenhouse effect would develop.

38. Asteroidal and cometary collision rate
    - If this rate were greater, too many species would become extinct.
    - If this rate were lesser, the Earth's crust would be too depleted of the materials essential for life.

39. Mass of body colliding with primordial Earth
    - If the body were smaller, Earth's atmosphere would have been too thick and the moon would have been too small.
    - If the body were greater, Earth's orbit and form would have been too greatly disturbed.

40. Timing of body colliding with primordial Earth
    - If the collision had occurred earlier, Earth's atmosphere would be too thick and the moon would be too small.
    - If the collision had occurred later, the Earth's atmosphere would be too thin and thus the sun would be too luminous for advanced life.

41. Location of body colliding with primordial Earth
    - If the body had just grazed the Earth, there would have been insufficient debris to form a large moon. Furthermore, the collision would have been inadequate to annihilate Earth's primordial atmosphere. Also, there would have been inadequate transfer of heavy elements to Earth.
    - If the body had collided too close to dead center, damage from the collision would have destroyed necessary conditions for (future) life.

42. Oxygen-to-nitrogen ratio in atmosphere
    - If this ratio were larger, advanced life functions would proceed too quickly.
    - If this ratio were smaller, advanced life functions would proceed too slowly.

43. Carbon dioxide level in atmosphere
    - If the level were greater, a runaway greenhouse effect would develop.
    - If the level were lesser, plants would be unable to maintain efficient photosynthesis.

44. Water vapor level in atmosphere
    - If the Earth's water vapor level were greater, a runaway greenhouse effect would develop.
    - If the Earth's water vapor level were smaller, rainfall would be too meager for advanced life on the land.

45. Atmospheric electric discharge rate
    - If the discharge rate were greater, too much fire destruction would occur.
    - If the discharge rate were smaller, too little nitrogen would be fixed in the atmosphere.

46. Ozone level in atmosphere
    - If the ozone level were greater, surface temperatures would be too low and there would be too little ultraviolet radiation for plant survival.
    - If the ozone level were lesser, surface temperatures would be too high and there would be too much ultraviolet radiation for plant survival.

47. Oxygen quantity in atmosphere
    - If the oxygen quantity were greater, plants and hydrocarbons would burn up too easily.
    - If the oxygen quantity were lesser, advanced animals would have too little oxygen to breathe.

48. Ratio of $^{40}$K, $^{235,238}$U, $^{232}$Th to iron
    - If this ratio were too low, there would be inadequate levels of plate tectonic and volcanic activity.
    - If this ratio were too high, the levels of radiation, earthquakes, and volcanoes would be too high for advanced life.

49. Rate of interior heat loss
    - If the rate were too low, there would be inadequate energy to drive the required levels of plate tectonic and volcanic activity.
    - If the rate were too high, plate tectonic and volcanic activity would shut down too quickly.

50. Seismic activity
    - If seismic activity were greater, too many life-forms would be destroyed.
    - If seismic activity were lesser, nutrients on the ocean floors from river runoff would not be recycled to continents through tectonics. Furthermore, not enough carbon dioxide would be released from carbonates.

51. Volcanic activity
    - If volcanic activity were lower, insufficient amounts of carbon dioxide and water vapor would be returned to the atmosphere. Also, soil mineralization would become too degraded for life.
    - If volcanic activity were higher, advanced life would be destroyed.

52. Rate of decline in tectonic activity
    - If the rate were slower, advanced life could never survive on Earth.
    - If the rate were faster, advanced life could never survive on Earth.

53. Rate of decline in volcanic activity
    - If the rate were slower, advanced life could never survive on Earth.
    - If the rate were faster, advanced life could never survive on Earth.

APPENDIX A: *Fine-Tuning for Life on Earth* 197

54. Timing of birth of continent formation
    - If the formation had begun too early, the silicate-carbonate cycle would have been destabilized.
    - If the formation had begun too late, the silicate-carbonate cycle would have been destabilized.

55. Oceans-to-continents ratio
    - If the ratio were greater, diversity and complexity of life-forms would be limited and the silicate-carbonate cycle would be destabilized.
    - If the ratio were smaller, diversity and complexity of life-forms would be limited and the silicate-carbonate cycle would be destabilized.

56. Rate of change in oceans-to-continents ratio
    - If the rate was slower, advanced life would lack the needed landmass area.
    - If the rate was faster, advanced life would be destroyed by the radical changes.

57. Global distribution of continents
    - If the continents were located too much in the southern hemisphere, seasonal differences would be too severe for advanced life.

58. Frequency and extent of ice ages
    - If these were smaller, insufficient fertile, wide, and well-watered valleys would have been produced for diverse and advanced life-forms. Also, there would be insufficient mineral concentrations for diverse and advanced life.
    - If these were greater, Earth would experience runaway freezing.

59. Soil mineralization
    - If the Earth's soil were too nutrient-poor, the diversity and complexity of life-forms would be limited.
    - If the Earth's soil were too nutrient-rich, the diversity and complexity of life-forms would be limited.

60. Gravitational interaction with a moon
    - If the gravitational interaction were greater, tidal effects on the oceans, atmosphere, and rotational period would be too severe.
    - If the gravitational interaction were lesser, orbital obliquity changes would cause climatic instabilities. Movement of nutrients and life from the oceans to the continents and vice versa would be insufficient. Also, the magnetic field would be too weak.

61. Jupiter distance
    - If the distance from Earth to Jupiter were greater, too many asteroid and comet collisions would occur on Earth.
    - If the distance from Earth to Jupiter were lesser, Earth's orbit would be unstable.

62. Jupiter mass
    - If Jupiter's mass were greater, Earth's orbit would be unstable.
    - If Jupiter's mass were lesser, too many asteroid and comet collisions would occur on Earth.

63. Drift in major planet distances
    - If the planet drift were greater, Earth's orbit would become unstable.
    - If the planet drift were lesser, too many asteroid and comet collisions would occur on Earth.

64. Major planet eccentricities
    - If the eccentricities of the major planets in this solar system were greater, Earth would be pulled out of the life support zone.

65. Major planet orbital instabilities
    - If orbital instabilities were greater, Earth's orbit would be pulled out of the life support zone.

66. Atmospheric pressure
    - If atmospheric pressure on Earth were too slight, liquid water would evaporate too easily and condense too infrequently. Additionally, weather and climate variation would be too

APPENDIX A: *Fine-Tuning for Life on Earth* 199

extreme and lungs could not function.

- If atmospheric pressure on Earth were too great, liquid water would not evaporate easily enough for land life. Also, insufficient sunlight and ultraviolet radiation would reach the planet's surface. There would be insufficient climate and weather variation. And lungs would not function.

67. Atmospheric transparency
    - If atmospheric transparency were lesser, an insufficient range of wavelengths of solar radiation would reach Earth's surface.
    - If atmospheric transparency were greater, too broad a range of wavelengths of solar radiation would reach Earth's surface.

68. Magnitude and duration of the sunspot cycle
    - If the magnitude of the cycle were lesser or the duration briefer, there would be insufficient variation in climate and weather.
    - If the magnitude of the cycle were greater or the duration longer, variation in climate and weather would be too great.

69. Continental relief
    - If the relief were smaller, there would be insufficient variation in climate and weather.
    - If the relief were greater, variation in climate and weather would be too great.

70. Chlorine quantity in atmosphere
    - If there were less chlorine, erosion rates, the acidity of rivers, lakes, and soils, and certain metabolic rates would all be insufficient for most life-forms.
    - If there were more chlorine, erosion rates, the acidity of rivers, lakes, and soils, and certain metabolic rates would be too high for most life-forms.

71. Iron quantity in oceans and soils
    - If there were less iron, the quantity and diversity of life would be too limited to support advanced life. And if the quantity

were very small, no life would be possible.
- If there were more iron, iron poisoning of at least advanced life would result.

72. Tropospheric ozone quantity
    - If there were less tropospheric ozone, insufficient cleansing of biochemical smogs would result.
    - If there were more tropospheric ozone, the respiratory failure of advanced animals, reduced crop yields, and the destruction of ozone-sensitive species would result.

73. Stratospheric ozone quantity
    - If there were less stratospheric ozone, too much ultraviolet radiation would reach the Earth's surface, causing skin cancers and reducing plant growth.
    - If there were more stratospheric ozone, too little ultraviolet radiation would reach the Earth's surface, causing reduced plant growth and insufficient vitamin production for animals.

74. Mesospheric ozone quantity
    - If there were less mesospheric ozone, circulation and chemistry of mesospheric gases would be so disturbed as to upset relative abundances of life-essential gases in the lower atmosphere.
    - If there were more mesospheric ozone, circulation and chemistry of mesospheric gases would be so disturbed as to upset relative abundances of life-essential gases in the lower atmosphere.

75. Quantity and extent of forest and grass fires
    - If these were lesser, growth inhibitors in the soils would accumulate. Soil nitrification would be insufficient. Also, there would be insufficient charcoal production for adequate soil water retention and absorption of certain growth inhibitors.
    - If these were greater, too many plant and animal life-forms would be destroyed.

76. Quantity of soil sulfur
    - If there were less sulfur in the soil, plants would become

deficient in certain proteins and die.
- If there were more sulfur in the soil, plants would die from sulfur toxins. The acidity of water and soil would become too great for life. Also, nitrogen cycles would be disturbed.

77. Biomass-to-comet-infall ratio
    - If this ratio were smaller, greenhouse gases would accumulate, triggering runaway surface temperature increase.
    - If this ratio were larger, greenhouse gases would decline, triggering a runaway freeze.

APPENDIX B

# PROBABILITIES FOR LIFE ON EARTH

This appendix presents an estimate for the probability of attaining the necessary parameters for life support on a planet. It includes 153 known parameters.[1]

| Parameter | Probability that feature will fall in the required range for physical life |
|---|---|
| local abundance and distribution of dark matter | .1 |
| galaxy cluster size | .1 |
| galaxy cluster location | .1 |
| galaxy size | .1 |
| galaxy type | .1 |
| galaxy mass distribution | .2 |
| galaxy location | .1 |
| variability of local dwarf galaxy absorption rate | .1 |
| star location relative to galactic center | .2 |
| star distance from corotation circle of galaxy | .005 |
| star distance from closest spiral arm | .1 |
| z-axis extremes of star's orbit | .02 |
| proximity of solar nebula to a type I supernova eruption | .01 |
| timing of solar nebula formation relative to type I supernova eruption | .01 |
| proximity of solar nebula to a type II supernova eruption | .01 |
| timing of solar nebula formation relative to type II supernova eruption | .01 |

| | |
|---|---:|
| flux of cosmic ray protons | .1 |
| variability of cosmic ray proton flux | .1 |
| number of stars in birthing cluster | .01 |
| star formation history in parent star vicinity | .1 |
| birth date of the star-planetary system | .01 |
| number of stars in system | .7 |
| number and timing of close encounters by nearby stars | .01 |
| proximity of close stellar encounters | .1 |
| masses of close stellar encounters | .1 |
| star age | .4 |
| star metallicity | .05 |
| ratio of $^{40}$K, $^{235,238}$U, $^{232}$Th to iron in star-planetary system | .02 |
| star orbital eccentricity | .1 |
| star mass | .001 |
| star luminosity change relative to speciation types and rates | .00001 |
| star color | .4 |
| star magnetic field | .1 |
| star magnetic field variability | .1 |
| stellar wind strength and variability | .1 |
| short period variation in parent star diameter | .3 |
| star's carbon-to-oxygen ratio | .01 |
| star's space velocity relative to Local Standard of Rest | .05 |
| star's short-term luminosity variability | .05 |
| star's long-term luminosity variability | .05 |
| amplitude and duration of star spot cycle | .1 |
| number and timing of solar system encounters with interstellar gas clouds | .1 |
| galactic tidal forces on planetary system | .2 |
| H3+ production | .1 |
| supernovae rates and locations | .01 |
| white dwarf binary types, rates, and locations | .01 |
| structure of comet cloud surrounding planetary system | .3 |
| planetary distance from star | .001 |
| inclination of planetary orbit | .5 |
| axis tilt of planet | .3 |
| rate of change of axial tilt | .01 |
| period and size of axial tilt variation | .1 |

# APPENDIX B: *Probabilities for Life on Earth*   205

| | |
|---|---|
| planetary rotation period | .1 |
| rate of change in planetary rotation period | .05 |
| planetary revolution period | .2 |
| planetary orbit eccentricity | .3 |
| rate of change of planetary orbital eccentricity | .1 |
| rate of change of planetary inclination | .5 |
| period and size of eccentricity variation | .1 |
| period and size of inclination variation | .1 |
| number of moons | .2 |
| mass and distance of moon | .01 |
| surface gravity (escape velocity) | .001 |
| tidal force from sun and moon | .1 |
| magnetic field | .01 |
| rate of change and character of change in magnetic field | .1 |
| albedo (planet reflectivity) | .1 |
| density | .1 |
| reducing strength of planet's primordial mantle | .3 |
| thickness of crust | .01 |
| timing of birth of continent formation | .1 |
| oceans-to-continents ratio | .2 |
| rate of change in oceans-to-continents ratio | .1 |
| global distribution of continents | .3 |
| frequency, timing, and extent of ice ages | .1 |
| frequency, timing, and extent of global snowball events | .1 |
| asteroidal and cometary collision rate | .1 |
| change in asteroidal and cometary collision rates | .1 |
| rate of change in asteroidal and cometary collision rates | .1 |
| mass of body colliding with primordial Earth | .002 |
| timing of body colliding with primordial Earth | .05 |
| location of body's collision with primordial Earth | .05 |
| position and mass of Jupiter relative to Earth | .01 |
| major planet eccentricities | .1 |
| major planet orbital instabilities | .05 |
| drift and rate of drift in major planet distances | .05 |
| number and distribution of planets | .01 |
| distance of gas giant planets from mean motion resonances | .02 |
| atmospheric transparency | .01 |

| | |
|---|---:|
| atmospheric pressure | .01 |
| atmospheric viscosity | .1 |
| atmospheric electric discharge rate | .01 |
| atmospheric temperature gradient | .01 |
| carbon dioxide level in atmosphere | .01 |
| rate of change in carbon dioxide level in atmosphere | .1 |
| rate of change in water vapor level in atmosphere | .01 |
| rate of change in methane level in early atmosphere | .01 |
| oxygen quantity in atmosphere | .01 |
| nitrogen quantity in atmosphere | .01 |
| chlorine quantity in atmosphere | .1 |
| carbon monoxide quantity in atmosphere | .1 |
| cobalt quantity in crust | .1 |
| arsenic quantity in crust | .1 |
| copper quantity in crust | .1 |
| boron quantity in crust | .1 |
| fluorine quantity in crust | .1 |
| iodine quantity in crust | .1 |
| manganese quantity in crust | .1 |
| nickel quantity in crust | .1 |
| phosphorus quantity in crust | .1 |
| tin quantity in crust | .1 |
| zinc quantity in crust | .1 |
| molybdenum quantity in crust | .05 |
| vanadium quantity in crust | .1 |
| chromium quantity in crust | .1 |
| selenium quantity in crust | .1 |
| iron quantity in oceans | .1 |
| tropospheric ozone quantity | .01 |
| stratospheric ozone quantity | .01 |
| mesospheric ozone quantity | .01 |
| water vapor level in atmosphere | .01 |
| oxygen-to-nitrogen ratio in atmosphere | .1 |
| quantity of greenhouse gases in atmosphere | .01 |
| rate of change in greenhouse gases in atmosphere | .01 |
| quantity of forest and grass fires | .01 |
| quantity of sea salt aerosols | .1 |

APPENDIX B: *Probabilites for Life on Earth*  207

| | |
|---|---|
| soil mineralization | .1 |
| quantity of anaerobic bacteria in the oceans | .01 |
| quantity of aerobic bacteria in the oceans | .01 |
| quantity, variety, and timing of sulfate-reducing bacteria | .001 |
| quantity of decomposer bacteria in soil | .01 |
| quantity of mycorrhizal fungi in soil | .01 |
| quantity of nitrifying microbes in soil | .01 |
| quantity and timing of vascular plant introductions | .001 |
| quantity, timing, and placement of carbonate-producing animals | .00001 |
| quantity, timing, and placement of methanogens | .00001 |
| quantity of soil sulfur | .1 |
| rate of interior heat loss for planet | .01 |
| quantity of sulfur in the planet's core | .1 |
| quantity of silicon in the planet's core | .1 |
| quantity of water at subduction zones in the crust | .01 |
| quantity of high-pressure ice in subducting crustal slabs | .1 |
| hydration rate of subducted minerals | .1 |
| tectonic activity | .05 |
| rate of decline in tectonic activity | .1 |
| volcanic activity | .1 |
| rate of decline in volcanic activity | .1 |
| continental relief | .1 |
| viscosity at Earth core boundaries | .01 |
| viscosity of lithosphere | .2 |
| biomass-to-comet-infall ratio | .01 |
| regularity of cometary infall | .1 |
| number, intensity, and location of hurricanes | .02 |
| dependency factors estimate | $10^{30}$ |
| longevity requirements estimate | $10^{13}$ |

The probability of a planet anywhere in the universe fitting within all 153 parameters is approximately $10^{-194}$. The maximum possible number of planets in the universe is estimated to be $10^{22}$. Thus, less than 1 chance in $10^{172}$ (100 thousand trillion trillion trillion trillion trillion trillion trillion trillion trillion trillion trillion trillion trillion) exists that even one such planet would occur anywhere in the universe.

APPENDIX C

# Fine-Tuning for Life in the Universe

For life to be possible in the universe, several characteristics must take on specific values, and these are listed below. In the case of several of these characteristics, and given the intricacy of their relationships, the indication of fine-tuning seems incontrovertible.[1]

1. Strong nuclear force constant
2. Weak nuclear force constant
3. Gravitational force constant
4. Electromagnetic force constant
5. Ratio of electromagnetic force constant to gravitational force constant
6. Ratio of proton to electron mass
7. Ratio of number of protons to number of electrons
8. Expansion rate of the universe
9. Mass density of the universe
10. Baryon (proton and neutron) density of the universe
11. Space energy density of the universe
12. Entropy level of the universe
13. Velocity of light
14. Age of the universe
15. Uniformity of radiation
16. Homogeneity of the universe
17. Average distance between galaxies
18. Average distance between stars
19. Average size and distribution of galaxy clusters
20. Fine structure constant
21. Decay rate of protons
22. Ground state energy level for helium-4
23. Carbon-12 to oxygen-16 nuclear energy level ratio

24. Decay rate for beryllium-8
25. Ratio of neutron mass to proton mass
26. Initial excess of nucleons over antinucleons
27. Polarity of the water molecule
28. Epoch for supernova eruptions
29. Frequency of supernova eruptions
30. Epoch for white dwarf binaries
31. Density of white dwarf binaries
32. Ratio of exotic matter to ordinary matter
33. Number of effective dimensions in the early universe
34. Number of effective dimensions in the present universe
35. Mass of the neutrino
36. Magnitude of big bang ripples
37. Size of the relativistic dilation factor
38. Magnitude of the Heisenberg uncertainty

Visit reasons.org/fine-tuning for a catalog of characteristics of the universe and Earth that require fine-tuning for life's existence, including relevant citations to the scientific literature.

# NOTES

**Chapter 1: The UFO Craze**
1. Richard Williams, ed., UFO: *The Continuing Enigma* (Pleasantville, NY: Readers Digest Association, 1991), 18–29.
2. For an informative overview of the UFO age, see Jerome Clark, "The UFO Phenomenon: A Historical Overview," *The UFO Encyclopedia*, 2nd ed. 1 (Detroit: Omnigraphics, 1998), x–xiv.
3. Paul Devereux and Peter Brookesmith, *UFOs and Ufology: The First Fifty Years* (New York: Facts on File, 1997), 12–25.
4. Devereux and Brookesmith, 21–23.
5. Quoted in John Spencer, ed., *The UFO Encyclopedia* (New York: Avon Books, 1991), s.v. "Arnold, Kenneth."
6. Clark, *UFO Encyclopedia*, s.v. "Arnold sighting."
7. Clark, "UFO Phenomenon," x.
8. Clark, *UFO Encyclopedia*, s.v. "University of Colorado UFO Project."
9. Clark, "UFO Phenomenon," xiii.
10. Clark, "A Note on UFO Terminology," xv.
11. UFO researcher John A. Saliba's categories have influenced my own. See John Saliba, "Religious Dimensions of UFO Phenomena," in *The Gods Have Landed*, ed. James R. Lewis (New York: State University of New York Press, 1995), 16–26.
12. David Kestenbaum, "Panel Says Some UFO Reports Worthy of Study," *Science* 281(1998), 21.
13. For an argument in support of a government cover-up, see Lawrence Fawcett and Barry J. Greenwood, *The UFO Cover-Up: What the Government Won't Say* (New York: Prentice-Hall, 1984). For a more sober view, see Saliba, "Religions Dimensions of UFO Phenomena," 21; and Mark Clark's two chapters in this volume (chapters 7 and 8).
14. For extensive information on various UFO research groups,

see Clark, *UFO Encyclopedia*, vols. 1 and 2.
15. Saliba, "Religious Dimensions of UFO Phenomena," 19–20.
16. Douglas Groothuis, *Confronting the New Age* (Downers Grove, IL: Inter Varsity, 1988), 29–30.

**Chapter 2: Types of UFOs**
1. CNI News, "Gallup Poll Indicates Strong Belief in Extraterrestrial Life," http://www.exosci.com/ufo, accessed January 6, 1998.
2. ABCNews.com: Roswell. "We Think the Truth Is Out There, Says Poll, http://archive.abcnews.com/sections/scitech, accessed July 2, 1998.
3. J. Allen Hynek, *The UFO Experience: A Scientific Inquiry* (New York: Marlow, 1998), 22.
4. Jacques Vallée, *Dimensions: A Casebook of Alien Contact* (New York: Ballantine Books, 1988), 231.
5. "Frequently Asked Questions About UFOs," Center for UFO Studies, http://www.cufos.org/FAQ_index.html, accessed June 12, 2001.
6. "Alien Contact: Interview with Timothy Good," *Rutherford* 5, no. 10 (October 1996), 17.
7. "Frequently Asked Questions," Center for UFO Studies.
8. John A. Saliba, "UFO Contactee Phenomena from a Sociopsychological Perspective: A Review," in *The Gods Have Landed*, ed. James R. Lewis (New York: State University of New York Press, 1995), 207–8.
9. Saliba, "Religious Dimensions of UFO Phenomena," in The Gods Have Landed, ed. James R. Lewis (New York: State University of New York Press, 1995), 207–8.
10. Ibid.
11. Hynek, *UFO Experience*, 33–16.
12. Jacques Vallée, *Confrontations: A Scientist's Search for Alien Contact* (New York: Ballantine Books, 1990), 211; John Ankerberg and John Weldon, *The Facts on UFOs and Other Supernatural Phenomena* (Eugene, OR: Harvest House, 1992), 7–8.

13. Vallée, *Confrontations*, 216–19.
14. John Spencer, ed., *The UFO Encyclopedia* (New York: Avon Books, 1991), s.v. "identified flying objects (IFOs)."
15. "Frequently Asked Questions," Center for UFO Studies.
16. *UFO: The Continuing Enigma* (Pleasantville, NY: Readers Digest Association, 1991), 136–39; *UFOs...The Mystery Resolved*, prod. Hugh Ross, Reasons To Believe, videocassette.
17. *UFO: The Continuing Enigma*, 129; *UFOs...The Mystery Resolved*.
18. Spencer, s.v. "hoax reports;" Jerome Clark, *The UFO Encyclopedia*, 2nd ed. 1 (Detroit: Omnigraphics, 1998), s.v. "hoaxes."
19. "Frequently Asked Questions," Center for UFO Studies.
20. Saliba, "Religious Dimensions of UFO Phenomena," 230–31.
21. Ibid., 241.
22. *UFO: The Continuing Enigma*, 133; Clark, s.v. "earthlights and tectonic strain theory."
23. *Encyclopedia Americana*, s.v. "unidentified flying object."
24. Philip J. Klass, *UFOs Explained* (New York: Random House, 1974); Philip J. Klass, *UFOs: The Public Deceived* (Buffalo: Prometheus Books, 1983); Philip J. Klass, *Bringing UFOs down to Earth* (Amherst, NY: Prometheus Books, 1997).
25. "UFOs: Are We Alone? The Truth Behind UFO Sightings," Nova, 1982, videocassette.
26. Quoted in Saliba, "Religious Dimensions of UFO Phenomena," 210.
27. Carl G. Jung, *Flying Saucers: The Myth of Things Seen in the Skies* (London: Kegan Paul, 1958).
28. Clark, s.v. "extraterrestrial hypothesis and ufology."
29. Quoted in Clark, s.v. "Friedman, Stanton Terry."
30. Vallée, *Dimensions*, 228–41; Clark, s.v. "extraterrestrial hypothesis and ufology."
31. Vallée, *Dimensions*, 228–41; Clark, s.v. "extraterrestrial hypothesis and ufology;" Ankerberg and Weldon, 11–13.
32. *UFOs...The Mystery Resolved*; William M. Alnor, UFOs in the New Age (Grand Rapids, MI: Baker, 1992), 80–81.
33. Ankerberg and Weldon, 16–27; Clark, s.v. "paranormal and

occult theories about UFOs;" *UFOs...The Mystery Resolved.*
34. Ankerberg and Weldon, 16–27; *UFOs...The Mystery Resolved*; Clifford Wilson and John Weldon, *Close Encounters: A Better Explanation* (San Diego: Master Books, 1978).
35. Vallée, *Dimensions*, 253.
36. Ankerberg and Weldon, 16–27; Clark, s.v. "paranormal and occult theories."

**Chapter 3: Life on Other Planets**
1. A. Wolszczan and D. A. Frail, "A Planetary System Around the Millisecond Pulsar PSR 1257+12," *Nature* 255 (1992): 145; A. Wolszczan, "Confirmation of Earth-Mass Planets Orbiting the Millisecond Pulsar PSR B1257+12," *Science* 264 (1994): 538.
2. Michel Major and Didier Queloz, "A Jupiter-Mass Companion to a Solar-Type Star," *Nature* 378 (1995), 355–59.
3. Jean Schneider, Extra Solar Planets Catalog, http://www.obspm.fr/encycl/catalog.html. This website is frequently updated.
4. Guillermo Gonzalez, "The Stellar Metallicity-Giant Planet Connection," *Monthly Notices of the Royal Astronomical Society* 285 (1997): 403–12; Guillermo Gonzalez, "Spectroscopic Analysis of the Parent Stars of Extrasolar Planetary System Candidates," *Astronomy and Astrophysics* 334 (1998): 221–38; Guillermo Gonzalez, George Wallerstein, and Steven H. Saar, "Parent Stars of Extrasolar Planets. IV. 14 Herculis, HD 187123, and HD 210277," *Astrophysical Journal Letters* 511 (1999): L111–L114; Guillermo Gonzalez, "Are Stars with Planets Anomalous?" *Monthly Notices of the Royal Astronomical Society* 308 (1999): 447–58; Guillermo Gonzalez and Chris Laws, "Parent Stars of Extrasolar Planets. V. HD 75289," *Astronomical Journal* 119 (2000): 390–96; Guillermo Gonzalez et al., "Parent Stars of Extrasolar Planets. VI. Abundance Analyses of 20 New Systems," *Astronomical Journal* 121 (2001): 432–52.
5. Guillermo Gonzalez, 2000, private communication.
6. S. Sahipal et al., "A Stellar Origin for the Short-Lived Nu-

clides in the Early Solar System," *Nature* 391 (1998): 559–661; G. J. Wasserburg, R. Gallino, and M. Busso, "A Test of the Supernova Trigger Hypothesis with 60Fe and 26Al," *Astrophysical Journal Letters* 500 (1998): L189–L193; Peter Hoppe et al., "Type II Supernova Matter in a Silicon Carbide Grain from the Murchison Meteorite," *Science* 272 (1996): 1314–16.
7. Ronald L. Gilliland et al., "A Lack of Planets in 47 Tucanae from a Hubble Space Telescope Search," *Astrophysical Journal Letters* 545 (2000): L47–L51.
8. Iosef S. Shklovskii and Carl Sagan, *Intelligent Life in the Universe* (San Francisco: Holden-Day, 1966), 343–50.
9. Shklovskii and Sagan, 409–13.
10. Robert H. Dicke, "Dirac's Cosmology and Mach's Principle," *Nature* 192 (1961): 440.
11. Hugh Ross, *The Creator and the Cosmos*, 2nd ed. (Colorado Springs, CO: NavPress, 1995), 132–44.
12. J. F. Kasting, D. P. Whitmire, and R. T. Reynolds, "Habitable Zones Around Main Sequence Stars," *Icarus* 101 (1993): 108–28; Darren M. Williams, James F. Kasting, and Richard Wade, "Habitable Moons Around Extrasolar Giant Planets," *Nature* 385 (1977): 234–36; Darren M. Williams, "The Stability of Habitable Planetary Environments" (Ph.D. thesis, Pennsylvannia State University, 1998), 111–20.
13. S. Vogt, G. Marcy, P. Butler, and K. Apps, "Six New Planets from the Keck Precision Velocity Survey," *Astrophysical Journal* 536 (2000): 902–14; Gonzalez, Wallerstein, and Saar, L111–L114; Ing-Guey Jiang and Wing-Huen Ip, "The Planetary System of Upsilon Andromedae," *Astronomy and Astrophysics* 367 (2001): 943–48; Jean Schneider, Extra Solar Planets Catalog, http://www.obspm.fr/encycl/catalog.html [Nov. 5, 2001].
14. Frederic A. Rasio and Eric B. Ford, "Dynamical Instabilities and the Formation of Extrasolar Planetary Systems," *Science* 274 (1996): 954–58; N. Murray et al., "Migrating Planets," Science 279 (1998): 69–72; D. N. C. Lin, P. Bodenheimer, and D. C. Richardson, "Orbital Migration of the Planetary

Companion of 51 Pegasi to Its Present Location," *Nature* 380 (1996): 606-7; Stuart J. Widenschilling and Francesco Marsari, "Gravitational Scattering as a Possible Origin for Giant Planets at Small Stellar Distances," *Nature* 384 (1996): 619-21; Gregory Laughlin and Fred C. Adams, "The Modification of Planetary Orbits in Dense Open Clusters," *Astrophysical Journal Letters* 508 (1998): L171-L174; Stuart Ross Taylor, Destiny or Chance: Our Solar System and Its Place in the Cosmos (Cambridge: Cambridge University Press, 1998).
15. Schneider.
16. William B. McKinnon, "Galileo at Jupiter—Meetings with Remarkable Moons," *Nature* 391 (1997): 23-26.
17. McKinnon, 24.
18. Hugh Ross, "A Look at the Case for Moon Life," *Facts & Faith* 12, fourth quarter 1998, 1-2; D. A. Gurnett et al., "Evidence for a Magnetosphere at Ganymede from Plasma-Wave Observations by the Galileo Spacecraft," *Nature* 384 (1996): 535-37.
19. M. G. Kivelson et al., "Discovery of Ganymede's Magnetic Field by the Galileo Spacecraft," *Nature* 384 (1996): 537-41.
20. Kivelson et al., 541.
21. Williams, "Stability of Habitable Planetary Environments," 115-17.
22. Ross, *Creator and the Cosmos*, 134-35.
23. Hugh Ross, "The Case for Creation Grows Stronger," *Facts & Faith* 4, first quarter 1990, 1-3; Tyler Volk and David Schwartzman, "Biotic Enhancement of Weathering and the Habitability of Earth," *Nature* 340 (1989): 457-60.
24. Katherine L. Moulton and Robert A. Berner, "Quantification of the Effect of Plants on Weathering: Studies in Iceland," *Geology* 26 (1998): 895-98.
25. David Stevenson, "Life-Sustaining Planets in Interstellar Space?" Nature 400 (1999): 32; Christopher Wills and Jeffrey Bada, *The Spark of Life* (Cambridge, MA: Perseus, 2000), 250-52.

## Chapter 4: Evolution's Probabilities

1. Richard Dawkins, *Climbing Mount Improbable* (New York: Norton, 1996).
2. Christopher Chyba and Carl Sagan, "Endogenous Production, Exogenous Delivery and Impact-Shock Synthesis of Organic Molecules: An Inventory for the Origins of Life," *Nature* 355 (1992): 125–32.
3. Richard A. Kerr, "Beating Up on a Young Earth, and Possibly Life," *Science* 290 (2000): 1677; B. A. Cohen, T. D. Swindle, and D. A. Kring, "Support for the Lunar Cataclysm Hypothesis from Lunar Meteorite Impact Melt Ages," *Science* 290 (2000): 1754–56.
4. S. J. Mojzsis et al., "Evidence for Life on Earth Before 3,800 Million Years Ago," *Nature* 384 (1996): 53–59.
5. Kevin A. Maher and David J. Stevenson, "Impact Frustration of the Origin of Life," *Nature* 331, 1988, 612–614; Verne R. Oberbeck and Guy Fogleman, "Impacts and the Origin of Life," *Nature* 339 (1989): 434; Norman H. Sleep et al., "Annihilation of Ecosystems by Large Asteroid Impacts on the Early Earth," *Nature* 342 (1989): 139–42.
6. Maher and Stevenson, 612–14.
7. Daniel P. Glavin et al., "Amino Acids in Martian Meteorite Nakla," *Book of Abstracts* (Ninth Meeting of the International Society for the Study of the Origin of Life, University of California, San Diego, July 11–16, 1999), 62; Keith A. Kvenvolden, "Chirality of Amino Acids in the Murchison Meteorite—A Historical Perspective," *Book of Abstracts* (Ninth Meeting of the International Society for the Study of the Origin of Life, University of California, San Diego, July 11–16, 1999), 40.
8. An example is Alton Biggs, Biology: *The Dynamics of Life* (Columbus: Glencoe/McGraw-Hill, 2000).
9. Michael H. Carr, *Water on Mars* (New York: Oxford University Press, 1996), 197–203; Michael C. Malin and Michael H. Carr, "Groundwater Formation of Martian Valleys," *Nature* 397 (1999): 590–91; M. C. Malin and Kenneth S. Edgett, "Evidence for Recent Groundwater and Surface

Runoff on Mars," *Science* 288 (2000): 2330–35; Michael H. Carr, "The Habitability of Early Mars," *Book of Abstracts* (Ninth Meeting of the International Society for the Study of the Origin of Life, University of California, San Diego, July 11–16, 1999), 61.

10. Malin and Edgett, 2330; Hugh Ross, *The Creator and the Cosmos*, 3rd ed. (Colorado Springs, CO: NavPress, 2001), 209–11.
11. Robert Shapiro, "The Homopolymer Problem in the Origin of Life," *Book of Abstracts* (Ninth Meeting of the International Society for the Study of the Origin of Life, University of California, San Diego, July 11–16, 1999), 48.
12. Fred Hoyle and N. C. Wickramasinghe, *Evolution from Space: A Theory of Cosmic Creationism* (New York: Simon & Schuster, 1981).
13. Paul Parsons, "Dusting Off Panspermia," *Nature* 383 (1996): 221–22.
14. Ross, *Creator and the Cosmos*, 178–80, 188–98.
15. H. J. Melosh, "Astrobiology I: Thinking Big to the Nitty-Gritty," Lunar and Planetary Science Conference, Houston, Texas, March 12–16, 2001. A brief report of this discovery may also be found in Richard A. Kerr, "Rethinking Water on Mars and the Origin of Life," *Science* 292 (2001): 40.
16. Hugh Ross, "Molecular Mystery Fuels Faith," *Facts & Faith* 9, second quarter 1995, 3–5; Fazale R. Rana and Hugh Ross, "Life from the Heavens? Not This Way..." *Facts for Faith*, first quarter 2000, 12, 14–15; Stephen Mason, "Extraterrestrial Handedness," *Nature* 389 (1997): 804.
17. Robert Shapiro, *Origins: A Skeptic's Guide to the Creation of Life on Earth* (New York: Summit Books, 1986), 128.
18. Elizabeth Pennisi, "Microbial Genomes Come Tumbling In," *Science* 277 (1997) 1433; Colin Patterson, *Evolution*, 2nd ed. (Ithaca, NY: Comstock Publishing Associates, 1999), 23; Gerard Deckert et al., "The Complete Genome of the Hyperthermophilic Bacterium Aquifex Aeolicus," *Nature* 392 (1998): 353–58; Andreas Ruepp et al., "The Genome Sequence of the Thermoacidophilic Scavenger Thermoplas-

ma Acidophilum," *Nature* 407 (2000): 508–13; Carol J. Bult et al., "Complete Genome Sequence of the Methanogenic Archaeon, Methanococcus Jannaschii," *Science* 273 (1996): 1058–73.
19. Mitchell M. Waldrop, "Finding RNA Makes Proteins Gives 'RNA World' a Big Boost," *Science* 256 (1992): 1396–97.
20. Thomas R. Cech, "The Chemistry of Self-Splicing RNA and RNA Enzymes," *Science* 236 (1987): 1532–39.
21. Harry F. Noller, Veronita Hoffarth, and Ludwika Zimniak, "Unusual Resistance of Peptidyl Transferase to Protein Extraction Procedures," *Science* 256 (1992): 1416–19.
22. Joseph A. Piccirilli et al., "Aminoacyl Esterase Activity of the Tetrahymena Ribozyme," *Science* 256 (1992): 1420–24.
23. John Horgan, "In the Beginning," *Scientific American* 264, no. 2, February 1991, 119.
24. Robert Shapiro, "Prebiotic Ribose Synthesis: A Critical Analysis," *Origins of Life and Evolution of the Biosphere* 18 (1988): 71–85; Horgan, "In the Beginning," 119; Robert Shapiro, "Protometabolism: A Scenario for the Origin of Life," *American Scientist*, July–August 1992, 387.
25. Robert Irion, "Ocean Scientists Find Life, Warmth in the Seas: RNA Can't Take the Heat," *Science* 279 (1998): 1303.
26. Ibid.
27. Ibid.
28. John Horgan and Paul Hoffman, "Visions of the Twenty-First Century: Will There Be Anything Left to Discover?" *Time* 155, no. 14, April 10, 2000.

## Chapter 5: Interstellar Space Travel
1. Hugh Ross, *The Creator and the Cosmos*, 3rd ed. (Colorado Springs, CO: NavPress, 2001), 176–87.
2. This comes from the NASA catalog of the 2,613 known stars within 81.5 light-years of Earth. NASA, http://nstars.arc.nasa.gov, accessed June 14, 2001. Jean Schneider, Extra Solar Planets Catalog, http://www.obspm.fr/encycl/catalog.html, accessed June 14, 2001. This is a frequently updated website.

3. Christopher F. Chyba, "Life Beyond Mars," *Nature* 382 (1996): 577.
4. James F. Crow, "The Odds of Losing at Genetic Roulette," *Nature* 397 (1999): 293–94.
5. Adam Eyre-Walker and Peter D. Keightley, "High Genomic Deleterious Mutation Rates in Hominids," *Nature* 397 (1999): 344–47.
6. John W. Wright, ed., *The New York Times 2000 Almanac* (New York: Penguin Reference, 1999), 487.

**Chapter 6: RUFOs—The Unexplained UFOs**
1. J. Allen Hynek and Jacques Vallée, *The Edge of Reality: A Progress Report on Unidentified Flying Objects* (Chicago: Henry Regnery, 1975), 221; J. Allen Hynek, *The UFO Experience: A Scientific Inquiry* (New York: Marlowe, 1998), 256–60.
2. Hynek and Vallée, x–xi.
3. Alan Baker, *UFO Sightings* (New York: TV Books, 1999), 7.
4. Hynek and Vallée, 3, 31.
5. Jacques Vallée, *Confrontations: A Scientist's Search for Alien Contact* (New York: Ballantine Books, 1990), 14; Peter A. Sturrock, "An Analysis of the Condon Report on the Colorado UFO Project," *Journal of Scientific Exploration* 1, no. 1 (1987): 75–100; Richard F. Haines and Jacques F. Vallée, "Photo Analysis of an Aerial Disk over Costa Rica," *Journal of Scientific Exploration* 3, no. 2 (1989): 113–31; Richard F. Haines and Jacques F. Vallée, "Photo Analysis of an Aerial Disk over Costa Rica: New Evidence," *Journal of Scientific Exploration* 4, no. 1 (1990): 71–74; Michel C. L. Bounias, "Biochemical Traumatology as a Potent Tool for Identifying Actual Stresses Elicited by Unidentified Sources: Evidence for Plant Metabolic Disorders in Correlation with a UFO Landing," *Journal of Scientific Exploration* 4, no. 1 (1990): 1–18; Jacques F. Vallée, "Return to Trans-en-Provence," *Journal of Scientific Exploration* 4, no. 1 (1990): 19–25; Jean-Jacques Velasco, "Report on the Analysis of Anomalous Physical Traces: The 1981 Trans-en-Provence UFO Case," *Journal of Scientific Exploration* 4, no. 1 (1990):

27–48; Jacques F. Vallée, "Five Arguments Against the Extraterrestrial Origin of Unidentified Flying Objects," *Journal of Scientific Exploration* 4, no. 1 (1990): 105–17; William Bramley, "Can the UFO Extraterrestrial Hypothesis and Vallée Hypothesis Be Reconciled?" *Journal of Scientific Exploration* 6, no. 1 (1992): 3–9; Bruce Maccabee, "Analysis and Discussion of the May 18, 1992, UFO Sighting in Gulf Breeze, Florida," *Journal of Scientific Exploration* 7, no. 3 (1993): 241–57; Bruce Maccabee, "Strong Magnetic Field Detected Following a Sighting of an Unidentified Flying Object," *Journal of Scientific Exploration* 8, no. 3 (1994): 347–65; Peter A. Sturrock, "Report on a Survey of the Membership of the American Astronomical Society Concerning the UFO Problem: Part 1," *Journal of Scientific Exploration* 8, no. 1 (1994): 1–45; Peter A. Sturrock, "Report on a Survey of the Membership of the American Astronomical Society Concerning the UFO Problem: Part 3," *Journal of Scientific Exploration* 8, no. 3 (1994): 309–46; Peter A. Sturrock et al., "Physical Evidence Related to UFO Reports: The Proceedings of a Workshop Held at the Pocantico Conference Center, Tarrytown, New York, September 29–October 4, 1997," *Journal of Scientific Exploration* 12, no. 2 (1998): 179–29; Jacques F. Vallée, "Estimates of Optical Power Output in Six Cases of Unexplained Aerial Objects with Defined Luminosity Characteristics," *Journal of Scientific Exploration* 12, no. 3 (1998): 345–58; Jacques F. Vallée, "Physical Analyses in Ten Cases of Unexplained Aerial Objects with Material Samples," *Journal of Scientific Exploration* 12, no. 3 (1998): 359–75; Bruce Maccabee, "Atmosphere or UFO? A Response to the 1997 SSE Review Panel Report," *Journal of Scientific Exploration* 13, no. 3 (1999): 421–59; Michael D. Swords, "Clyde Tombaugh, Mars, and UFOs," *Journal of Scientific Exploration* 13, no. 4 (1999): 685–94; Ann Druffel, Robert M. Wood, and Eric Kelson, "Reanalysis of the 1965 Heflin UFO Photos," *Journal of Scientific Exploration* 14, no. 4 (2000): 583–622; Hynek, 11–15, 19–20, 22–85, 90–98, 100–6, 108, 110, 112, 115–37, 148–50; Hynek and Vallée, xii–xv,

4–8, 10–15, 18–19, 27, 29, 31, 34–42, 44–50, 59–62, 64–69, 86–87, 89–102, 129–38, 152–54, 159–64, 172–73, 258–60, 265–88, 291–92; Ted Phillips Jr., *Physical Traces Associated with UFO Sightings* (Evanston, IL: Center for UFO Studies, 1975); Baker, 15–148, 203–15, 271–91; Richard F. Haines, *Melbourne Episode: Case Study of a Missing Pilot* (Los Altos, CA: LDA Press, 1987); Richard H. Hall, *The UFO Evidence: A Thirty-Year Report* 2 (Lanham, MD: Scarecrow Press, 2001), 1–70, 73–112, 115–45, 149–66, 169–94, 198–238, 241–302, 315–20, 322–52, 409–32, 457–98; Jacques Vallée, *Dimensions: A Casebook of Alien Contact* (New York: Ballantine Books, 1988), xv–xvii, 6–64, 66–96, 231–32; Jacques Vallée, *Confrontations: A Scientist's Search for Alien Contact* (New York: Ballantine Books, 1990), 22–34, 36–56, 63–83, 91–100, 164–73, 180–97.
6. Hall, 261–66; Baker, 58–59; Phillips.
7. Hynek, 129.
8. Hall, 261.
9. Baker, 56–58.
10. Baker, 57, 66–68, 86–88; Vallée, *Confrontations*, 94–100; Vallée, "Physical Analyses in Ten Cases," 363–64, 366, 368; Sturrock et al., 202–4; Vallée, "Return to Trans-en-Provence," 19–25; Velasco, 27–48; Vallée, "Physical Analyses in Ten Cases," 359–75.
11. Hynek, 128–29, 131–33; Hall, 261–66; Vallée, *Confrontations*, 91–100; Baker, 67; Sturrock et al., 204–6; Bounias, 1–18; Vallée, "Return to Trans-en-Provence," 19–25; Velasco, 27–48.
12. Vallée, *Confrontations*, 92–93.
13. Sturrock et al., 205–6.
14. Sturrock et al., 204–5; Bounias, 1–18.
15. Hall, 278–82; Hynek and Vallée, 5, 268–79; Baker, 58.
16. Hynek, 110, 115, 121, 129; Hall, 266–72; Baker, 61–64, 91–92.
17. Vallée, *Confrontations*, 15, 101–25, 204–6; Baker, 63; Hynek and Vallée, 159–61; Haines, Melbourne Episode.
18. Vallée, *Confrontations*, 15.

19. Hall, 247–60, 272–78; Mark Rodeghier, *UFO Reports Involving Vehicle Interference* (Evanston, IL: Center for UFO Studies, 1981); Sturrock et al., 196–99; Hynek and Vallée, xiv–xv, 11, 51, 159; Hynek, 110, 115–28.
20. Sturrock et al., 197–99; Richard F. Haines, "Fifty-Six Aircraft Pilot Sightings Involving Electromagnetic Effects," *MUFON 1992 International UFO Symposium Proceedings* (Albuquerque, New Mexico, July 1992), 102–8.
21. Jennie Zeidman, *A Helicopter-UFO Encounter over Ohio* (Evanston, IL: Center for UFO Studies, 1979); Sturrock et al., 200–1.
22. Jerome Clark, *The UFO Encyclopedia*, 2nd ed. 1 (Detroit: Omnigraphics, 1998), 95–99.
23. Hynek and Vallée, 7–8.
24. John Spencer, ed., *The UFO Encyclopedia* (New York: Avon Books, 1991), 64.
25. Jerome Clark, *The UFO Encyclopedia*, 2nd ed. 2 (Detroit: Omnigraphics, 1998), 663–67; Richard C. Henry, "UFOs and NASA," *Journal of Scientific Exploration* 2, no. 2, 1988, 93–142.

## Chapter 7: Government Cover-Ups

1. For an excellent discussion of these and related stories, see Curtis Peebles, *Watch the Skies! A Chronicle of the Flying Saucer Myth* (Washington: Smithsonian Institution Press, 1994).
2. See, in order of publication, Charles Berlitz and William L. Moore, *The Roswell Incident* (New York: Berkley, 1980); Kevin D. Randle and Donald R. Schmitt, *UFO Crash at Roswell* (New York: Avon Books, 1991); Don Berliner and Stanton T. Friedman, *Crash at Corona* (New York: Paragon House, 1992); Kevin D. Randle and Donald R. Schmitt, *The Truth About the UFO Crash at Roswell* (New York: Avon Books, 1994); and Karl T. Pflock, "Roswell, the Air Force, and Us," *International UFO Reporter* 19, no. 6, 1994.
3. Charles A. Ziegler, chapters 1 and 2, in Benson Saler, Charles A. Ziegler, and Charles B. Moore, *UFO Crash at Roswell: The*

*Genesis of a Modern Myth* (Washington: Smithsonian Institution Press, 1997), 1–73.

4. Operation Majestic-12 was supposedly a group of a dozen top scientists, military officers, and intelligence specialists set up by President Harry Truman in 1947 to study the Roswell remains and the humanoid bodies that had been recovered nearby. See the reports on the Majestic-12 forgeries in Philip J. Klass, "Special Report: The MJ-12 Documents," *Skeptical Inquirer* 12, no. 2, 1988, 137–146; Philip J. Klass, "New Evidence of MJ-12 Hoax," *Skeptical Inquirer* 14, no. 2, 1990, 135–140; and Harold K. Haines, "CIA's Role in the Study of UFOs, 1947–1990," *Studies in Intelligence* 1, 1997, 87, fn. 93.

5. See Leon Jaroff, "Did Aliens Really Land?" *Time*, June 23, 1997, 68–71.

6. Headquarters, United States Air Force, *The Roswell Report: Fact vs. Fiction in the New Mexico Desert* (Washington: U.S. Government Printing Office, 1995), 7.

7. See Philip J. Klass, *The Real Roswell Crashed Saucer Cover-up* (New York: Prometheus Books, 1997), 13–23; and Headquarters, U.S. Air Force, *The Roswell Report: Case Closed* (Washington: U.S. Government Printing Office, 1997), 5–9.

8. The U.S. Air Force made the first official mention of the previously top secret Mogul project in its publication *The Roswell Report: Fact vs. Fiction in the New Mexico Desert.* However, Karl Pflock (cited above) discovered the Air Force's interest in the project earlier.

9. For an excellent discussion, see Charles B. Moore, "The Early New York University Balloon Flights," in Saler, Ziegler, and Moore, 74–114.

10. Saler, Ziegler, and Moore, 169–180.

11. Headquarters, U.S. Air Force, *Roswell Report: Fact vs. Fiction*, 3.

12. Saler, Ziegler, and Moore, 177.

13. Philip J. Klass proposes that the true cover-up is not that of U.S. involvement in UFO phenomena but that of the pro-UFO conspiracy theorists who stand to gain material

rewards for continuing to perpetuate a sham. See Klass, *Real Roswell Crashed Saucer Cover-up.*
14. Haines, 68; J. Allen Hynek, *The UFO Experience: A Scientific Inquiry* (New York: Ballantine Books, 1972), 199.
15. Haines, 68.
16. Keep in mind that this was the beginning of the McCarthy era, a domestic overreaction to several significant foreign policy losses to the Soviets, who drew an iron curtain around Eastern Europe in the late 1940s, exploded an atomic bomb in 1949, supported Mao Zedong's communist takeover in China in the same year, and encouraged North Korea's war on South Korea in 1950.
17. Haines, 72; Hynek, *UFO Experience*, 190–191.
18. Hynek, *UFO Experience*, 191.
19. Haines, 73.
20. Jacques Vallée, *Dimensions: A Casebook of Alien Contact* (New York: Ballantine Books, 1989), 209–210.
21. J. Allen Hynek, The Hynek UFO Report (New York: Dell, 1977), 12.
22. Vallée, 211.
23. Hynek, *Hynek UFO Report,* 12–13.
24. The Center for UFO Studies maintains an active presence on the Internet (http://www.cufos.org), in addition to supporting UFO researchers and publications, including some who hold to the pro-UFO conspiracy theory discussed above.
25. Edward U. Condon and Daniel S. Gilmor, eds., *Final Report of the Scientific Study of Unidentified Flying Objects* (New York: Bantam Books, 1969), 965.
26. See Office of Assistant Secretary of Defense, news release, "Air Force to Terminate Project Blue Book," December 17, 1969.
27. The classic study of bureaucratic politics is found in Graham T. Allison, *Essence of Decision: Explaining the Cuban Missile Crisis* (Boston: Little, Brown, 1971), 67–143. A contemporary use of organizational theory is found in Scott D. Sagan and Kenneth N. Waltz, *The Spread of Nuclear Weapons:*

*A Debate* (New York: Norton, 1995).
28. See David Darlington, *Area 51: The Dreamland Chronicles: The Legend of America's Most Secret Military Base* (New York: Holt, 1997), 14; and Federation of American Scientists, Intelligence Reform Project, "Aurora/Senior Citizen," http://www.fas.org/irp/mystery/aurora.htm, accessed June 12, 2001. See also Ben R. Rich and Leo Janos, *Skunk Work: A Personal Memoir of My Years at Lockheed* (New York: Little, Brown, 1994).
29. For the Web site of the Federation of American Scientists, see http://www.fas.org.
30. Martin Van Creveld, *Technology and War: From 2000 B.C. to the Present* (New York: Free Press, 1991).
31. Darlington, chaps. 3, 6.
32. "More Aurora," *Secrecy and Government Bulletin*, no. 30, 1994, http://www.fas.org.sgp/bulletin/sec30, accessed August 6, 1998.
33. See the comments of former astronaut Buzz Aldrin on the transcripts of the show *Politically Incorrect*, aired August 1, 1996, Aliens on Earth, http://www.aliensonearth.com/area51/articles/1996/kanipe_960808.html, accessed June 12, 2001.
34. See Department of Defense, *Soviet Acquisition of Militarily Significant Western Technology: An Update* (Washington: U.S. Government Printing Office, September 1985).
35. See Richard Bernstein and Ross H. Munro, *The Coming Conflict with China* (New York: Knopf, 1997); and John J. Fialka, *War by Other Means: Economic Espionage in America* (New York: Norton, 1997).

## Chapter 8: Government Conspiracies
1. Jerome Clark, "The UFO Phenomenon: A Historical Overview," *The UFO Encyclopedia*, 2nd ed. 1 (Detroit: Omnigraphics, 1998), x–xiv.
2. Beth A. Fischer, "Perception, Intelligence Errors, and the Cuban Missile Crisis," *Intelligence and National Security* 13, no. 3, 1998, 152.

3. See Robert Jervis, *Perception and Misperception in International Politics* (Princeton, N.J.: Princeton University Press, 1976).
4. Jervis, *Perception and Misperception*, 148.
5. Fischer, 153.
6. The best (but not the only) example is found in Timothy Good, *Above Top Secret: The Worldwide UFO Cover-up* (New York: William Morrow, 1988).
7. In the field of defense and foreign affairs, the magazine is humorously referred to as "Aviation Leak and Space Mythology" because of the richness of its information on technology development from highly classified documents. *The Washington Times* is well known for its reporting on numerous classified Pentagon and Central Intelligence Agency documents, much to the chagrin of the administration in power, whether Republican or Democrat.
8. See, for example, memorandum by William J. Clinton, "Unauthorized Disclosure of Classified Information," May 2, 1995, Federation of American Scientists, Project on Government Secrecy, http://www.fas.org/sgp/clinton/leaks, accessed February 7, 1998.
9. See, for example, Lawrence Fawcett and Barry J. Greenwood, *The UFO Cover-Up: What the Government Won't Say* (New York: Prentice-Hall, 1984).
10. Philip J. Klass, *The Real Roswell Crashed Saucer Cover-up* (New York: Prometheus Books, 1997), 208.
11. Gallup Poll, conducted September 3–5, 1996, cited in "Gallup Poll Indicates Strong Belief in Extraterrestrial Life," July 24, 1997, http://www.exosci.com/ufo/news/8, accessed January 6, 1998.
12. Rick Martin and T. Trent Gregory, "Conspiracy Mania Feeds Our Growing National Paranoia," *Newsweek*, January 6, 1997, http://www.ufomind.com/area51/articles/1996/newsweek, accessed August 6, 1998.
13. For a humorous example, see Leon Jaroff, "Did Aliens Really Land?" *Time*, June 23, 1997, http://pathfinder.com/time/magazine/1997/dom/970623/society, accessed January 8, 1998.

14. Robert Jervis, "Perception and Coping with Threats," in Robert Jervis, Richard Ned Lebow, and Janice Gross Stein, *Psychology and Deterrence* (Baltimore: Johns Hopkins University Press, 1985), 300–301.

## Chapter 9: Nature and Supernature

1. Robert Jastrow, *God and the Astronomers*, 2nd ed. (New York: Norton, 1992), 7–149; Hugh Ross, *The Fingerprint of God*, 2nd ed. (Orange, CA: Promise, 1991), 27–118.
2. Roger Penrose, "An Analysis of the Structure of Space-Time" (Adams Prize Essay, 1966, Cambridge University); Stephen W. Hawking, "Singularities and the Geometry of Space-Time" (Adams Prize Essay, 1966, Cambridge University); Stephen W. Hawking and George F. R. Ellis, "The Cosmic Black-Body Radiation and the Existence of Singularities in our Universe," *Astrophysical Journal* 152 (1968): 25–36.
3. Stephen W. Hawking and Roger Penrose, "The Singularities of Gravitational Collapse and Cosmology," *Proceedings of the Royal Society of London* A 314 (1970): 529–48; Stephen W. Hawking and George F. R. Ellis, *The Large-Scale Structure of Space-Time* (Cambridge: Cambridge University Press, 1973).
4. J. H. Taylor et al., "Experimental Constraints on Strong-field Relativistic Gravity," *Nature* 355 (1992), 132–36; Roger Penrose, *Shadows of the Mind* (New York: Oxford University Press, 1994), 230; An excellent and fully derived description of the Lense-Thirring effect and of the experimental difficulties in its confirmation may be found in Hans C. Ohanian and Remo Ruffini, *Gravitation and Space time*, 2nd ed. (New York: Norton, 1994), 220–28, 419–23; Ignazio Ciufolini et al., "Test of General Relativity and Measurement of the Lense-Thirring Effect with Two Earth Satellites," *Science* 279 (1998): 2100–3; Ron Cowen, "Einstein's General Relativity: It's a Drag," *Science News* 152 (1997): 308; David Kestenbaum, "X-ray Flickers Reveal a Space Warp," Science 280 (1998): 674–75; Luigi Stella and Mario Vietri, "Lense-Thirring Precession and Quasi-Periodic Oscillations

in Low-Mass X-Ray Binaries," *Astrophysical Journal Letters* 492 (1998): L59–L62; Peter G. Jonker, Mariano Méndez, and Michiel van der Klis, "Discovery of a New Third Kilohertz Quasi-Periodic Oscillation in 4U 1608–52, 4U 1728–34, and 4U 1636–53: Sidebands to the Lower Kilohertz Quasi-Periodic Oscillation?" *Astrophysical Journal Letters* 540 (2000): L29–L32; G. S. Bisnovatyi-Kogan, "At the Border of Eternity," Science 279 (1998): 1321; Andrew Watson, "Einstein's Theory Rings True," *Science* 280 (1998): 205.
5. Jacob D. Bekenstein, "Nonsingular General-Relativistic Cosmologies," *Physical Review* D 11 (1975): 2072–75; Leonard Parker and Yi Wang, "Avoidance of Singularities in Relativity Through Two-Body Interactions," *Physical Review* D 42 (1990): 1877–83; Arvind Borde, "Open and Closed Universes, Initial Singularities, and Inflation," *Physical Review* D 50 (1994): 3692–3702; Arvind Borde and Alexander Vilenkin, "Eternal Inflation and the Initial Singularity," *Physical Review Letters* 72 (1994): 3305–8; Arvind Borde and Alexander Vilenkin, "Violation of the Weak Energy Condition in Inflating Spacetimes," *Physical Review* D 56 (1997): 717–23.
6. A review of the discoveries and evidence for ten-dimensional cosmology is provided in Hugh Ross, *Beyond the Cosmos*, 2nd ed. (Colorado Springs, CO: NavPress, 1999), 27–46.
7. Hugh Ross, *The Creator and the Cosmos*, 3rd ed. (Colorado Springs, CO: NavPress, 2001), 145–67.
8. Fazale R. Rana and Micah Lott, "Hume vs. Paley: These 'Motors' Settle the Debate," *Facts for Faith*, second quarter 2000, 34–39. See also Stephen M. Block, "Real Engines of Creation," *Nature* 386 (1997): 217.
9. Ross, *Creator and the Cosmos*, 161.
10. Ibid., 152–57, 176–98.
11. Hugh Ross, *The Genesis Question* (Colorado Springs, CO: NavPress, 1998), 51–57, 64–65, 149–51; Hugh Ross, "Mutations Exceed Expectations," *Connections* 1, second quarter 1999, 3; Adam Eyre-Walker and Peter D. Keightley, "High Genomic Deleterious Mutation Rates in Hominids," *Nature* 397 (1999): 344–47.

12. Ross, *The Genesis Question*, 151–54; Hugh Ross, "Bacteria Help Prepare Earth for Life," *Connections* 3, no. 1, 2001, 4; Crisogono Vasconcelos and Judith A. McKenzie, "Sulfate Reducers—Dominant Players in a Low-Oxygen World," *Science* 290, 2000, 1711–1712; Matthias Labrenz et al., "Formation of Sphalerite (ZnS) Deposits in Natural Biofilms of Sulfate-Reducing Bacteria," *Science* 290 (2000): 1744–47.
13. Hugh Ross, "Anthropic Principle: A Precise Plan for Humanity," *Facts for Faith*, first quarter 2002, 24–31.

**Chapter 10: The Interdimensional Hypothesis**
1. Hugh Ross, *Beyond the Cosmos*, 2nd ed. (Colorado Springs, CO: NavPress, 1999), 34–46.
2. Jacques Vallée, *Dimensions: A Casebook of Alien Contact* (New York: Ballantine Books, 1988), 253. Vallée's term "multiverse" means something quite different from the standard cosmological definition. The multiverse concept in current cosmology speculates that the universe is an expanding space-time "bubble" that erupted from a larger seething quantum foam of space-time fluctuations.
3. Hugh Ross, *The Creator and the Cosmos*, 3rd ed. (Colorado Springs, CO: NavPress, 2001), 127–36, 169–74.
4. Ross, *Creator and the Cosmos*, 23–258.
5. Ross, *Creator and the Cosmos*, 77–118; William F. Campbell, *The Qur'an and the Bible in the Light of History and Science* (Upper Darb, PA: Middle East Resources, 1992); Dean Halverson, "Urantia: The Brotherhood Book," *Spiritual Counterfeits Project Newsletter*, August 1981, 1, 3–5.

**Chapter 11: A Closer Look at RUFOs**
1. David Kestenbaum, "Panel Says Some UFO Reports Worthy of Study," *Science* 281 (1998): 21; J. Allen Hynek and Jacques Vallée, *The Edge of Reality: A Progress Report on Unidentified Flying Objects* (Chicago: Henry Regnery, 1975); J. Allen Hynek, *The UFO Experience: A Scientific Inquiry* (New York: Marlowe, 1998); Richard H. Hall, *A Thirty-Year Report 2 of The UFO Evidence* (Lanham, MD: Scarecrow Press,

2001); Alan Baker, *UFO Sightings* (New York: TV Books, 1999); Jacques Vallée, *Dimensions: A Casebook of Alien Contact* (New York: Ballantine Books, 1988); Jacques Vallée, *Confrontations: A Scientist's Search for Alien Contact* (New York: Ballantine Books, 1990); Peter A. Sturrock, "An Analysis of the Condon Report on the Colorado UFO Project," *Journal of Scientific Exploration* 1, no. 1 (1987): 75–100; Peter A. Sturrock, "Report on a Survey of the Membership of the American Astronomical Society Concerning the UFO Problem: Part 1," *Journal of Scientific Exploration* 8, no. 1 (1994): 1–45; Peter A. Sturrock, "Report on a Survey of the Membership of the American Astronomical Society Concerning the UFO Problem: Part 3," *Journal of Scientific Exploration* 8, no. 3 (1994): 309–46; Peter A. Sturrock et al., "Physical Evidence Related to UFO Reports: The Proceedings of a Workshop Held at the Pocantico Conference Center, Tarrytown, New York, September 29–October 4, 1997," *Journal of Scientific Exploration* 12, no. 2 (1998): 179–229; Jacques F. Vallée, "Physical Analyses in Ten Cases of Unexplained Aerial Objects with Material Samples," *Journal of Scientific Exploration* 12, no. 3 (1998): 359–75; Bruce Maccabee, "Atmosphere or UFO? A Response to the 1997 SSE Review Panel Report," *Journal of Scientific Exploration* 13, no. 3 (1999): 421–59; Ted Phillips Jr., *Physical Traces Associated with UFO Sightings* (Evanston, IL: Center for UFO Studies, 1975); Mark Rodeghier, *UFO Reports Involving Vehicle Interference* (Evanston, IL: Center for UFO Studies, 1981); Richard F. Haines, "Fifty-Six Aircraft Pilot Sightings Involving Electromagnetic Effects," *MUFON 1992 International UFO Symposium Proceedings* (Albuquerque, New Mexico, July 1992), 102–8.
2. Vallée, *Confrontations*, 14.
3. David Michael Jacobs, *The UFO Controversy in America* (Bloomington: Indiana University Press, 1975), 5–34.
4. Jacobs, 36–37.
5. Vallée, *Dimensions*, xvi–96.
6. Vallée, *Dimensions*, xvi–96.

7. Vallée, *Dimensions*, 76–78.
8. Vallée, *Confrontations*, 118–124; Hall, 346.
9. Sturrock, "Report on a Survey: Part 1," 1–45; Sturrock, "Report on a Survey: Part 3," 309–46.
10. John Spencer, ed., *The UFO Encyclopedia* (New York: Avon Books, 1991), 253–54.
11. Vallée, *Dimensions*, x.
12. Vallée, *Dimensions*, x.
13. John A. Keel, *UFOs: Operation Trojan Horse* (New York: Putnam, 1970), 143.
14. Hynek, 12.
15. Hynek, 115, 121; Hall, 266–72.
16. Vallée, *Dimensions*, xvi.
17. Vallée, *Dimensions*, 253.
18. Hynek and Vallée, 262.
19. Hynek and Vallée, 258.
20. Paul Davies, *Are We Alone?* (New York: Basic Books, 1995), 133.
21. Keel, 215.
22. Lynn G. Catoe, *UFOs and Related Subjects: An Annotated Bibliography Prepared for the U. S. Air Force Office of Scientific Research* (Washington: U.S. Government Printing Office, n.d.), introduction.

**Chapter 12: Abductees**
1. Jerome Clark, *The UFO Encyclopedia*, 2nd ed. 1 (Detroit: Omnigraphics, 1998), s.v. "abduction phenomenon." Thomas E. Bullard's extensive article in this encyclopedia provides excellent background information on all phases of the abduction phenomenon and reflects both personal insight and objectivity. For these reasons, this researcher quotes and cites him extensively.
2. Clark, s.v. "abduction phenomenon."
3. Clark, s.v. "abduction phenomenon."
4. Clark, s.v. "abduction phenomenon."
5. John Spencer, ed., *The UFO Encyclopedia* (New York: Avon Books, 1991), s.v. "Hill, Betty and Barney."
6. Clark, s.v. "Hill abduction case"; William A. Alnor, *UFOs in*

the *New Age* (Grand Rapids, MI: Baker, 1992), 78.
7. Clark, s.v. "Hill abduction case."
8. Clark, s.v. "abduction phenomenon."
9. Spencer, s.v. "Walton, Travis"; Clark, s.v. "Walton abduction case."
10. Clark, s.v. "Budd Hopkins."
11. Clark, s.v. "abduction phenomenon."
12. Clark, s.v. "abduction phenomenon."
13. Clark, s.v. "abduction phenomenon."
14. Clark, s.v. "abduction phenomenon"; see also John Whitmore, "Religious Dimensions of the UFO Abductee Experience," in *The Gods Have Landed*, ed. James R. Lewis (New York: State University of New York Press, 1995), 69–74.
15. Jacques Vallée, *Revelations: Alien Contact and Human Deception* (New York: Ballantine Books, 1991), 286.
16. Whitmore, 65–84.
17. Clark, s.v. "abduction phenomenon."
18. See Budd Hopkins, *Intruders: The Incredible Visitations at Copley Woods* (New York: Random House, 1987); and Budd Hopkins, *Missing Time: A Documented Study of UFO Abductions* (New York: Richard Marek, 1981).
19. Clark, s.v. "abduction phenomenon."
20. Clark, s.v. "abduction phenomenon."
21. Clark, s.v. "abduction phenomenon."
22. Clark, s.v. "abduction phenomenon."
23. Whitmore, 66.
24. Whitmore, 75.
25. Whitmore, 75.
26. Clark, s.v. "abduction phenomenon."
27. Spencer, s.v. "hypnotic regression"; Whitmore, 66; UFO: *The Continuing Enigma* (Pleasantville, NY: Readers Digest Association, 1991), 81.
28. Clark, s.v. "abduction phenomenon."
29. Clark, s.v. "abduction phenomenon."
30. Clark, s.v. "abduction phenomenon"; see also Whitmore, 75.
31. Clark, s.v. "abduction phenomenon."
32. Clark, s.v. "abduction phenomenon."
33. Clark, s.v. "abduction phenomenon."

34. Clark, s.v. "abduction phenomenon."
35. Clark, s.v. "abduction phenomenon."
36. John A. Saliba, "Religious Dimensions of UFO Phenomena," in *The Gods Have Landed*, ed. James R. Lewis (New York: State University of New York Press, 1995), 25.
37. Jacques Vallée, *Dimensions: A Casebook of Alien Contact* (New York: Ballantine Books, 1988), 242–51.
38. See Hopkins, Intruders; Hopkins, Missing Time; and David M. Jacobs, *Secret Life: Firsthand Accounts of UFO Abductions* (New York: Simon & Schuster, 1992).
39. Clark, s.v. "abduction phenomenon."
40. Clark, s.v. "abduction phenomenon."
41. Clark, s.v. "abduction phenomenon."
42. Clark, s.v. "abduction phenomenon."
43. UFO: The Continuing Enigma, 84–85, 100–5.
44. See *UFO: The Continuing Enigma*, 105, 122, 133–34; and Michael A. Persinger, "The Tectonic Strain Theory as an Explanation for UFO Phenomena: A Non-Technical Review of the Research, 1970–1990," *Journal of UFO Studies* 2, new series (1990): 105–37.
45. See Philip J. Klass, *UFO Abductions: A Dangerous Game* (Buffalo: Prometheus Books, 1988).
46. See *UFO: The Continuing Enigma*, 84; and Robert A. Baker and Joe Nickell, *Missing Pieces: How to Investigate Ghosts, UFOs, Psychics, and Other Mysteries* (Buffalo: Prometheus Books, 1992).
47. *UFO: The Continuing Enigma*, 105.
48. Clark, s.v. "abduction phenomenon."
49. See Hopkins, *Intruders*; Hopkins, *Missing Time*; and Jacobs, *Secret Life*.
50. Clark, s.v. "paranormal and occult theories about UFOs."
51. "A Dark Door to the Occult: An Interview with John Weldon," *Rutherford* 5, no. 10, (1996): 19.

**Chapter 13: Contactees**
1. J. Gordon Melton and George M. Eberhart, "The Flying Saucer Contactee Movement, 1950–1994: A Bibliography," in

*The Gods Have Landed*, ed. James R. Lewis (New York: State University of New York Press, 1995), 252.
2. J. Gordon Melton, "The Contactees: A Survey," in *Gods Have Landed*, 2-7.
3. Jerome Clark, *The UFO Encyclopedia*, 2nd ed. 1 (Detroit: Omnigraphics, 1998), s.v. "contactees."
4. Melton, 4.
5. Melton, 6; Clark, s.v. "contactees."
6. Ron Rhodes, *New Age Movement* (Grand Rapids, Mich.: Zondervan, 1995), 25; Melton, 6.
7. Melton, 6.
8. Rhodes, 26.
9. Melton, 7.
10. Elliot Miller, "Channeling Spiritistic Revelations for the New Age," part 1, *Christian Research Journal* 11, 1987; William M. Alnor, *UFOs in the New Age* (Grand Rapids, Mich.: Baker, 1992), 196-198. For further background and evaluation of *The Urantia Book*, see Eric Pement, "Urantia: The Great Cult Mystery," *Christian Research Journal*, Fall 1996, 48-49.
11. Melton, 7-8.
12. Christian apologist and specialist on New Age beliefs Elliot Miller defines channeling as "the practice of attempting communication with departed human or extra-human intelligences (usually nonphysical) through the agency of a human medium, with the intent of receiving paranormal information and/or having direct experience of metaphysical realities." Elliot Miller, *A Crash Course on the New Age Movement* (Grand Rapids, Mich.: Baker, 1989), 141. The term "channeling," which is used widely in New Age circles, actually came into use from contactees.
13. Clark, s.v. "Adamski, George."
14. Alnor, 87-89.
15. Quoted in *UFO: The Continuing Enigma* (Pleasantville, N.Y.: Reader's Digest Association, 1991), 72.
16. Alnor, 88.
17. Clark, s.v. "Adamski, George."

18. Clark, s.v. "Adamski, George."
19. Melton and Eberhart, 259.
20. Alnor, 89–90; Clark, s.v. "Van Tassel, George W."
21. Clark, s.v. "Van Tassel, George W."
22. Clark, s.v. "contactees."
23. Clark, s.v. "Van Tassel, George W."; Alnor, 89.
24. Clark, s.v. "contactees."
25. Clark, s.v. "contactees."
26. *UFO: The Continuing Enigma*, 75.
27. John A. Saliba, "Religious Dimensions of UFO Phenomena," in *The Gods Have Landed*, ed. James R. Lewis (New York: State University of New York Press, 1995), 25.
28. Clark, s.v. "contactees."
29. Robert S. Ellwood and Harry B. Partin, *Religious and Spiritual Groups in Modern America*, 2nd ed. (Upper Saddle River, NJ: Prentice-Hall, 1988), 111.
30. Melton and Eberhart, 252.
31. Quoted in Clark, s.v. "contactees."

**Chapter 14: UFO Cults**
1. J. Gordon Melton, "The Contactees: A Survey," in *The Gods Have Landed*, ed. James R. Lewis (New York: State University of New York Press, 1995), 9.
2. William M. Alnor, "UFO Cults Are Flourishing in New Age Circles," *Christian Research Journal*, Summer 1990, 35.
3. Other UFO cults operative today include the Aetherius Society; Amalgamated Flying Saucer Clubs of America; Ashtar Command; the Association of Sananda and Sanat Kumara; the Cosmic Circle of Fellowship; the Cosmic Star Temple; Deval UFO, Inc.; the George Adamski Foundation; the Jesusonian Foundation (Urantia); Last Day Messengers; Mark-Age, Inc.; One World Family; the Raelian movement; the Semjase Silver Star Center; the Solar Cross Foundation; the Solar Light Retreat; Unarius; the Universariun Foundation; the Universe Society Church Science of Life (UNISOC); White Star; and World Understanding. See J. Gordon Melton, *Encyclopedia of American Religions*, 4th ed. (Detroit:

Gale Research, 1992), 727; Alnor, 35.
4. John A. Saliba, "Religious Dimensions of UFO Phenomena," in *Gods Have Landed*, 27.
5. Saliba, 27.
6. *UFO: The Continuing Enigma* (Pleasantville, N.Y.: Readers Digest Association, 1991), 107.
7. Robert S. Ellwood and Harry B. Partin, *Religious and Spiritual Groups in Modern America*, 2nd. ed. (Upper Saddle River, N.J.: Prentice-Hall, 1988), 114.
8. Jerome Clark, *The UFO Encyclopedia*, 2nd ed. 1 (Detroit: Omnigraphics, 1998), s.v. "Aetherius Society."
9. Ellwood and Partin, 126.
10. Ellwood and Partin, 126–127.
11. Melton, 677.
12. *UFO: The Continuing Enigma*, 111–113; Melton, 683.
13. Clark, s.v. "Unarius Academy of Science."
14. Melton, 683.
15. Melton, 683.
16. Clark, s.v. "Unarius Academy of Science."
17. Clark, s.v. "Heaven's Gate."
18. Robert W. Balch, "Waiting for the Ships: Disillusionment and the Revitalization of Faith in Bo and Peeps's UFO Cult," in *Gods Have Landed*, 145.
19. Clark, s.v. "Heaven's Gate."
20. Balch, 143.
21. Clark, s.v. "Heaven's Gate."
22. Clark, s.v. "Heaven's Gate."
23. Clark, s.v. "Heaven's Gate."
24. Susan Jean Palmer, "Women in the Raelian Movement: New Religious Experiments in Gender and Authority," in *Gods Have Landed*, 106–14.
25. Palmer, 106–14.
26. Alnor, 35; Palmer, 106.
27. Palmer, 110.
28. Palmer, 107.
29. Palmer, 107.
30. Palmer, 107.

31. Palmer, 107.
32. Palmer, 112–113.
33. Alnor, 35; Palmer, 107.
34. Ronald H. Nash, *Faith and Reason* (Grand Rapids, MI: Zondervan, 1988), 24.
35. For further study of the topic of worldview thinking, especially from a Christian perspective, see Ronald H. Nash, *Worldviews in Conflict* (Grand Rapids, MI: Zondervan, 1992), 16–72; James W. Sire, *The Universe Next Door* (Downers Grove, IL: InterVarsity, 1988); William J. Wainwright, *Philosophy of Religion* (Belmont, CA: Wadsworth Publishing, 1988), 166–75.

## Chapter 15: The Bible and UFOs

1. Hugh Ross, *The Christmas Star* (Pasadena, CA: Reasons To Believe, 1990).
2. C. S. Lewis, "Dogma and the Universe," in *The Grand Miracle and Other Essays on Theology and Ethics from God in the Dock*, ed. Walter Hooper (New York: Ballantine Books, 1990), 14.
3. Stephen W. Hawking, *A Brief History of Time* (New York: Bantam Books, 1988), 126–27.
4. J. Gordon Melton, "The Contactees: A Survey," in *The Gods Have Landed*, ed. James R. Lewis (New York: State University of New York Press, 1995), 7–8.
5. Michael Green, *Exposing the Prince of Darkness* (Ann Arbor, MI: Servant, 1991), 118–24.
6. Ron Rhodes, *New Age Movement* (Grand Rapids, MI: Zondervan, 1995), 41.
7. According to the Aetherius Society, Jesus Christ is a Venusian. See Clark, s.v. "Aetherius Society." *The Urantia Book* presents Jesus as extraterrestrial and radically alters His nature, teaching, and mission. See Clifford Wilson and John Weldon, *Close Encounters: A Better Explanation* (San Diego, CA: Master Books, 1978), 233–41.
8. To see just how well the revelations of the contactees match with occultism and New Age thinking, see Wilson and Wel-

don, 311–19; and Elliot Miller, *A Crash Course on the New Age Movement* (Grand Rapids, MI: Baker, 1989).
9. The Spiritual Counterfeits Project (P.O. Box 4308, Berkeley, CA 94704) has published several articles on UFOs. Also, Walter R. Martin, the late president of the Christian Research Institute (P.O. Box 7000, Rancho Santa Margarita, CA 92688), produced a set of twelve audiotapes called *World of the Occult.*

## Appendix A: Fine-Tuning for Life on Earth

*Appendix A is based on research found in many different resources. Rather than identify each factual source with numbers scattered throughout the appendix, the following are relevant sources arranged in approximately the order in which they relate to the material in the appendix:*

1. R. E. Davies and R. H. Koch, "All the Observed Universe Has Contributed to Life," *Philosophical Transactions of the Royal Society of London* B 334 (1991): 391–403. Michael H. Hart, "Habitable Zones About Main Sequence Stars," *Icarus* 37 (1979): 351–57; William R. Ward, "Comments on the Long-Term Stability of the Earth's Oliquity," *Icarus* 50, 1982, 444–48; Carl D. Murray, "Seasoned Travellers," *Nature* 361 (1993): 586–87; Jacques Laskar and P. Robutel, "The Chaotic Obliquity of the Planets," *Nature* 361 (1993): 608–12; Jacques Laskar, F. Joutel, and P. Robutel, "Stabilization of the Earth's Obliquity by the Moon," *Nature* 361 (1993): 615–17; H. E. Newsom and S. R. Taylor, "Geochemical Implications of the Formation of the Moon by a Single Giant Impact," *Nature* 338 (1989): 29–34; W. M. Kaula, "Venus: A Contrast in Evolution to Earth," *Science* 247 (1990): 1191–96; Robert T. Rood and James S. Trefil, *Are We Alone? The Possibility of Extraterrestrial Civilizations* (New York: Scribner, 1983); John D. Barrow and Frank J. Tipler, *The Anthropic Cosmological Principle* (New York: Oxford University Press), 1986, 510–575; Don L. Anderson, "The Earth as a Planet: Paradigms and Paradoxes," *Science* 223 (1984): 347–55; I. H. Campbell and S. R. Taylor, "No

Water, No Granite—No Oceans, No Continents," *Geophysical Research Letters* 10 (1983): 1061–64; Brandon Carter, "The Anthropic Principle and Its Implications for Biological Evolution," *Philosophical Transactions of the Royal Society of London* A 310 (1983): 352–63; Allen H. Hammond, "The Uniqueness of the Earth's Climate," *Science* 187 (1975): 245; Owen B. Toon and Steve Olson, "The Warm Earth," *Science* 85 (1985): 50–57; George Gale, "The Anthropic Principle," *Scientific American* 245, no. 6 (1981): 154–71; Hugh Ross, *Genesis One: A Scientific Perspective* (Pasadena, CA: Reasons To Believe, 1983), 6–7; Ron Cottrell, *The Remarkable Spaceship Earth* (Denver: Accent Books, 1982); D. Ter Harr, "On the Origin of the Solar System," *Annual Review of Astronomy and Astrophysics* 5 (1967): 267–78; George Greenstein, *The Symbiotic Universe* (New York: William Morrow, 1988), 68–97; John M. Templeton, "God Reveals Himself in the Astronomical and in the Infinitesimal," *Journal of the American Scientific Affiliation* (1984): 196–98; Michael H. Hart, "The Evolution of the Atmosphere of the Earth," *Icarus* 33 (1978): 23–39; Tobias Owen, Robert D. Cess, and V. Ramanathan, "Enhanced $CO_2$ Greenhouse to Compensate for Reduced Solar Luminosity on Early Earth," *Nature* 277 (1979): 640–41; John Gribbin, "The Origin of Life: Earth's Lucky Break," *Science Digest* (May 1983): 36–102; P. J. E. Peebles and Joseph Silk, "A Cosmic Book of Phenomena," *Nature* 346 (1990): 233–39; Michael H. Hart, "Atmospheric Evolution, the Drake Equation, and DNA: Sparse Life in an Infinite Universe," in *Philosophical Cosmology and Philosophy*, ed. John Leslie (New York: Macmillan, 1990), 256–66; Stanley L. Jaki, *God and the Cosmologists* (Washington: Regnery Gateway, 1989), 177–84; R. Monastersky, "Speedy Spin Kept Early Earth from Freezing," *Science News* 143 (1993): 373; Editors, "Our Friend Jove," *Discover* (July 1993), 15; Jacques Laskar, "Large-Scale Chaos in the Solar System," *Astronomy and Astrophysics* 287 (1994): 109–13; Richard A. Kerr, "The Solar System's New Diversity," *Science* 265 (1994): 1360–62;

Richard A. Kerr, "When Comparative Planetology Hit Its Target," *Science* 265 (1994): 1361; W. R. Kuhn, J. C. G. Walker, and H. G. Marshall, "The Effect on Earth's Surface Temperature from Variations in Rotation Rate, Continent Formation, Solar Luminosity, and Carbon Dioxide," *Journal of Geophysical Research* 94 (1989): 11, 129-31, 136; Gregory S. Jenkins, Hal G. Marshall, and W. R. Kuhn, "Pre-Cambrian Climate: The Effects of Land Area and Earth's Rotation Rate," *Journal of Geophysical Research*, D 98 (1993): 8785-91; K. J. Zahnle and J. C. G. Walker, "A Constant Daylength During the Precambrian Era?" *Precambrian Research* 37 (1987): 95-105; M. J. Newman and R. T. Rood, "Implications of the Solar Evolution for the Earth's Early Atmosphere," *Science* 198 (1977): 1035-37; J. C. G. Walker and K. J. Zahnle, "Lunar Nodal Tides and Distance to the Moon During the Precambrian," *Nature* 320 (1986): 600-2; J. F. Kasting and J. B. Pollack, "Effects of High $CO_2$ Levels on Surface Temperatures and Atmospheric Oxidation State of the Early Earth," *Journal of Atmospheric Chemistry* 1 (1984): 403-28; H. G. Marshall, J. C. G. Walker, and W. R. Kuhn, "Long-Term Climate Change and the Geochemical Cycle of Carbon," *Journal of Geophysical Research* 93 (1988): 791-801; Pieter G. van Dokkum et al., "A High Merger Fraction in the Rich Cluster MS 1054-03 at z = 0.83: Direct Evidence for Hierarchical Formation of Massive Galaxies," *Astrophysical Journal Letters* 520 (1999): L95-L98; Anatoly Klypin, Andrey V. Kravtsov, and Octavio Valenzuela, "Where Are the Missing Galactic Satellites?" *Astrophysical Journal* 522 (1999): 82-92; Roland Buser, "The Formation and Early Evolution of the Milky Way Galaxy," *Science* 287 (2000): 69-74; Robert Irion, "A Crushing End for Our Galaxy," *Science* 287 (2000): 62-64; D. M. Murphy et al., "Influence of Sea Salt on Aerosol Radiative Properties in the Southern Ocean Marine Boundary Layer," *Nature* 392 (1998): 62-65; Neil F. Comins, *What If the Moon Didn't Exist?* (New York: HarperCollins, 1993), 2-8, 53-65; Hugh Ross, "Lunar Origin Update," *Facts & Faith* 9, first quarter 1995, 1-3; Jack

J. Lissauer, "It's Not Easy to Make the Moon," *Nature* 389 (1997): 327–28; Sigeru Ida, Robin M. Canup, and Glen R. Stewart, "Lunar Accretion from an Impact-Generated Disk," *Nature* 389 (1997): 353–57; Louis A. Codispoti, "The Limits to Growth," *Nature* 387 (1997): 237; Kenneth H. Coale, "A Massive Phytoplankton Bloom Induced by an Ecosystem-Scale Iron Fertilization Experiment in the Equatorial Pacific Ocean," *Nature* 383 (1996): 495–99; P. Jonathan Patchett, "Scum of the Earth After All," *Nature* 382 (1996): 758; William R. Ward, "Comments on the Long-Term Stability of the Earth's Obliquity," *Icarus* 50 (1982): 444–48; Carl D. Murray, "Seasoned Travellers," *Nature* 361 (1993): 586–87; Jacques Laskar and P. Robutel, "The Chaotic Obliquity of the Planets," *Nature* 361 (1993): 608–12; Jacques Laskar, F. Joutel, and P. Robutel, "Stabilization of the Earth's Obliquity by the Moon," *Nature* 361 (1993): 615–17; S. H. Rhie et al., "On Planetary Companions to the MACHO 98-BLG-35 Microlens Star," *Astrophysical Journal* 533 (2000): 378–91; Ron Cowen, "Less Massive Than Saturn?" *Science News* 157 (2000): 220–22; Hugh Ross, "Planet Quest—A Recent Success," *Connections* 2, second quarter 2000, 1–2; G. Gonzalez, "Spectroscopic Analyses of the Parent Stars of Extrasolar Planetary Systems," *Astronomy and Astrophysics* 334 (1998): 221–38; Guillermo Gonzalez, "New Planets Hurt Chances for ETI," *Facts & Faith* 12, fourth quarter 1998, 2–4; Editors, "The Vacant Interstellar Spaces," *Discover* (April 1996), 18, 21; Theodore P. Snow and Adolf N. Witt, "The Interstellar Carbon Budget and the Role of Carbon in Dust and Large Molecules," *Science* 270 (1995): 1455–57; Richard A. Kerr, "Revised Galileo Data Leave Jupiter Mysteriously Dry," *Science* 272 (1996): 814–15; Adam Burrows and Jonathan Lumine, "Astronomical Questions of Origin and Survival," *Nature* 378 (1995): 333; George Wetherill, "How Special Is Jupiter?" *Nature* 373 (1995): 470; B. Zuckerman, T. Forveille, and J. H. Kastner, "Inhibition of Giant-Planet Formation by Rapid Gas Depletion Around Young Stars," *Nature* 373 (1995): 494–96;

Hugh Ross, "Our Solar System, the Heavyweight Champion," *Facts & Faith* 10, second quarter 1996, 6; Guillermo Gonzalez, "Solar System Bounces in the Right Range for Life," *Facts & Faith* 11, first quarter 1997, 4–5; C. R. Brackenridge, "Terrestrial Paleoenvironmental Effects of a Late Quaternary-Age Supernova," *Icarus* 46 (1981): 81–93; M. A. Ruderman, "Possible Consequences of Nearby Supernova Explosions for Atmospheric Ozone and Terrestrial Life," *Science* 184 (1974): 1079–81; G. C. Reid et al., "Effects of Intense Stratospheric Ionization Events," *Nature* 275 (1978): 489–92; B. Edvardsson et al., "The Chemical Evolution of the Galactic Disk. I. Analysis and Results," *Astronomy & Astrophysics* 275 (1993): 101–52; J. J. Maltese et al., "Periodic Modulation of the Oort Cloud Comet Flux by the Adiabatically Changed Galactic Tide," *Icarus* 116 (1995): 255–68; Paul R. Renne et al., "Synchrony and Causal Relations Between Permian-Triassic Boundary Crisis and Siberian Flood Volcanism," *Science* 269 (1995): 1413–16; Hugh Ross, "Sparks in the Deep Freeze," *Facts & Faith* 11, first quarter 1997, 5–6; T. R. Gabella and T. Oka, "Detection of $H_3^+$ in Interstellar Space," *Nature* 384 (1996): 334–35; Hugh Ross, "Let There Be Air," *Facts & Faith* 10, third quarter 1996, 2–3; David J. DesMarais, Harold Strauss, Roger E. Summons, and J. M. Hayes, "Carbon Isotope Evidence for the Stepwise Oxidation of the Proterozoic Environment," *Nature* 359 (1992): 605–9; Donald E. Canfield and Andreas Teske, "Late Proterozoic Rise in Atmospheric Oxygen Concentration Inferred from Phylogenetic and Sulphur-Isotope Studies," *Nature* 382 (1996): 127–32; Alan Cromer, *Uncommon Sense: The Heretical Nature of Science* (New York: Oxford University Press, 1993), 175–76; Hugh Ross, "Drifting Giants Highlight Jupiter's Uniqueness," *Facts & Faith* 10, fourth quarter 1996, 4; Hugh Ross, "New Planets Raise Unwarranted Speculation About Life," *Facts & Faith* 10, first quarter 1996, 1–3; Hugh Ross, "Jupiter's Stability," *Facts & Faith* 8, third quarter 1994, 1–2; Christopher Chyba, "Life Beyond Mars," *Nature* 382 (1996): 577; E. Skindrad,

"Where Is Everybody?" *Science News* 150 (1996): 153; Stephen H. Schneider, *Laboratory Earth: The Planetary Gamble We Can't Afford to Lose* (New York: Basic Books, 1997), 25, 29–30; Guillermo Gonzalez, "Mini-Comets Write New Chapter in Earth-Science," *Facts & Faith* 11, third quarter 1997, 6–7; Miguel A. Goñi, Kathleen C. Ruttenberg, and Timothy I. Eglinton, "Sources and Contribution of Terrigenous Organic Carbon to Surface Sediments in the Gulf of Mexico," *Nature* 389 (1997): 275–78; Paul G. Falkowski, "Evolution of the Nitrogen Cycle and Its Influence on the Biological Sequestration of $CO_2$ in the Ocean," *Nature* 387 (1997): 272–74; John S. Lewis, *Physics and Chemistry of the Solar System* (San Diego: Academic Press, 1995), 485–92; Hugh Ross, "Earth Design Update: Ozone Times Three," *Facts & Faith* 11, fourth quarter 1997, 4–5; W. L. Chameides, P. S. Kasibhatla, J. Yienger, and H. Levy II, "Growth of Continental-Scale Metro-Agro-Plexes, Regional Ozone Pollution, and World Food Production," *Science* 264 (1994): 74–77; Paul Crutzen and Mark Lawrence, "Ozone Clouds over the Atlantic," *Nature* 388 (1997): 625; Paul Crutzen, "Mesospheric Mysteries," *Science* 277 (1997): 1951–52; M. E. Summers et al., "Implications of Satellite OH Observations for Middle Atmospheric $H_2O$ and Ozone," *Science* 277 (1997): 1967–70; K. Suhre et al., "Ozone-Rich Transients in the Upper Equatorial Atlantic Troposphere," *Nature* 388 (1997): 661–63; L. A. Frank, J. B. Sigwarth, and J. D. Craven, "On the Influx of Small Comets into the Earth's Upper Atmosphere. II. Interpretation," *Geophysical Research Letters* 13 (1986): 307–10; David Deming, "Extraterrestrial Accretion and Earth's Climate," *Geology*, in press; T. A. Muller and G. J. MacDonald, "Simultaneous Presence of Orbital Inclination and Eccentricity in Prozy Climate Records from Ocean Drilling Program Site 806," *Geology* 25 (1997): 3–6; Clare E. Reimers, "Feedback from the Sea Floor," *Nature* 391 (1998): 536–37; Hilairy E. Hartnett, Richard G. Keil, John I. Hedges, and Allan H. Devol, "Influence of Oxygen Exposure Time on Organic

Carbon Preservation in Continental Margin Sediments," *Nature* 391 (1998): 572–74; Tina Hesman, "Greenhouse Gassed: Carbon Dioxide Spells Indigestion for Food Chains," *Science News* 157 (2000): 200–2; Claire E. Reimers, "Feedbacks from the Sea Floor," *Nature* 391 (1998): 536–37; S. Sahijpal et al., "A Stellar Origin for the Short-Lived Nuclides in the Early Solar System," *Nature* 391 (1998): 559–61; Stuart Ross Taylor, *Destiny or Chance: Our Solar System and Its Place in the Cosmos* (New York: Cambridge University Press, 1998); Peter D. Ward and Donald Brownlee, *Rare Earth: Why Complex Life Is Uncommon in the Universe* (New York: Springer-Verlag, 2000); Dean L. Overman, *A Case Against Accident and Self-Organization* (New York: Rowman & Littlefield, 1997), 31–150; Michael J. Denton, *Nature's Destiny* (New York: Free Press,1998), 1–208; D. N. C. Lin, P. Bodenheimer, and D. C. Richardson, "Orbital Migration of the Planetary Companion of 51 Pegasi to Its Present Location," *Nature* 380 (1996): 606–7; Stuart J. Weidenschilling and Francesco Mazari, "Gravitational Scattering as a Possible Origin of Giant Planets at Small Stellar Distances," *Nature* 384 (1996): 619–21; Frederic A. Rasio and Eric B. Ford, "Dynamical Instabilities and the Formation of Extrasolar Planetary Systems," *Science* 274 (1996): 954–56; N. Murray, B. Hansen, M. Holman, and S. Tremaine, "Migrating Planets," *Science* 279 (1998): 69–72; Alister W. Graham, "An Investigation into the Prominence of Spiral Galaxy Bulges," *Astronomical Journal* 121 (2001): 820–40; Fred C. Adams and Gregory Laughlin, "Constraints on the Birth Aggregate of the Solar System," *Icarus* 150 (2001): 151–62; G. Bertelli and E. Nasi, "Star Formation History in the Solar Vicinity," *Astronomical Journal* 121 (2001): 1013–23; Nigel D. Marsh and Henrik Svensmark, "Low Cloud Properties Influenced by Cosmic Rays," *Physical Review Letters* 85 (2000): 5004–7; Gerhard Wagner et al., "Some Results Relevant to the Discussion of a Possible Link Between Cosmic Rays and the Earth's Climate," *Journal of Geophysical Research* 106 (2001): 3381–87; E. Pallé and C. J.

Butler, "The Influence of Cosmic Rays on Terrestrial Clouds and Global Warming," *Astronomy & Geophysics* 41 (2000): 4.19–4.22; B. Gladman and M. J. Duncan, "Fates of Minor Bodies in the Outer Solar System," *Astronomical Journal* 100 (1990): 1680–93; S. Alan Stern and Paul R. Weissman, "Rapid Collisional Evolution of Comets During the Formation of the Oort Cloud," *Nature* 409 (2001): 589–91; Christopher P. McKay and Margarita M. Marinova, "The Physics, Biology, and Environmental Ethics of Making Mar Habitable," *Astrobiology* 1 (2001): 89–109.

## Appendix B: Probabilities for Life on Earth

*All the references for appendix A apply to appendix B as well. What follow are additional relevant references for appendix B:*

1. Yu N. Mishurov and L. A. Zenina, "Yes, the Sun Is Located near the Corotation Circle," *Astronomy & Astrophysics* 341 (1999): 81–85; Guillermo Gonzalez, "Is the Sun Anomalous?" *Astronomy & Geophysics* 40, no. 5 (1999): 5.25–5.29; Ray White III and William C. Keel, "Direct Measurement of the Optical Depth in a Spiral Galaxy," *Nature* 359 (1992): 129–30; W. C. Keel and R. E. White III, "HST and ISO Mapping of Dust in Silhouetted Spiral Galaxies," *American Astronomical Society Meeting* 191, #75.01 (December 1997); Raymond E. White III, William C. Keel, and Christopher J. Conselice, "Seeing Galaxies Through Thick and Thin. I. Optical Opacity Measures in Overlapping Galaxies," *Astrophysical Journal* 542 (2000): 761–78; M. Emillio and J. R. Kuhn, "On the Constancy of the Solar Diameter," *Astrophysical Journal* 543 (2000): 1008–10; Douglas Gough, "Sizing Up the Sun," *Nature* 410 (2001): 313–14; John Vanermeer et al., "Hurricane Disturbance and Tropical Tree Species Diversity," *Science* 290 (2000): 788–91; Nicholas R. Bates, Anthony H. Knap, and Anthony F. Michaels, "Contribution of Hurricanes to Local and Global Estimates of Air-Sea Exchange of $CO_2$," *Nature* 395 (1998): 58–61; John Emsley, *The Elements,* 3rd ed. (Oxford: Clarendon, 1998), 24, 40,

56, 58, 60, 62, 78, 102, 106, 122, 130, 138, 152, 160, 188, 198, 214, 222, 230; Rob Rye, Phillip H. Kuo, and Heinrich D. Holland, "Atmospheric Carbon Dioxide Concentrations Before 2.2 Billion Years Ago," *Nature* 378 (1995): 603–5; Robert A. Muller and Gordon J. MacDonald, "Glacial Cycles and Orbital Inclination," Nature 377 (1995): 107–8; A. Evans, N. J. Beukes, and J. L. Kirschvink, "Low Latitude Glaciation in the Palaeoproterozoic Era," *Nature* 386 (1997): 262–66; Hugh Ross, "Rescued from Freeze-Up," *Facts & Faith* 11, second quarter 1997, 3; Hugh Ross, "New Developments in Martian Meteorite," *Facts & Faith* 10, fourth quarter 1996, 1–3; Paul Parsons, "Dusting Off Panspermia," *Nature* 383 (1996): 221–22; P. Jonathan Patchett, "Scum of the Earth After All," *Nature* 382 (1996): 758; Hubert P. Yockey, "The Soup's Not One," *Facts & Faith* 10, fourth quarter 1996, 10–11; M. Schlidowski, "A 3,800-Million-Year Isotopic Record of Life from Carbon in Sedimentary Rocks," *Nature* 333 (1988): 313–18; H. P. Yockey, *Information Theory and Molecular Biology* (Cambridge and New York: Cambridge University Press, 1992); C. De Duve, *Vital Dust* (New York: Basic Books, 1995). See also C. De Duve, *Blueprint for a Cell: The Nature and Origin of Life* (Burlington, NC: Neil Patterson, 1991); Hugh Ross, "Wild Fires Under Control," *Facts & Faith* 11, first quarter 1997, 1–2; Peter D. Moore, "Fire Damage Soils Our Forest," *Nature* 384 (1996): 312–13; A. U. Mallik, C. H. Gimingham, and A. A. Rahman, "Ecological Effects of Heather Burning I. Water Infiltration, Moisture Retention, and Porosity of Surface Soil," *Journal of Ecology* 72 (1984): 767–76; Hugh Ross, "Evidence for Fine-Tuning," *Facts & Faith* 11, second quarter 1997, 2. Herbert J. Kronzucker, M. Yaeesh Siddiqi, and Anthony D. M. Glass, "Conifer Root Discrimination Against Soil Nitrate and the Ecology of Forest Succession," *Nature* 385 (1997): 59–61; John M. Stark and Stephen C. Hart, "High Rates of Nitrification and Nitrate Turnover in Undisturbed Coniferous Forests," *Nature* 385 (1997): 61–64; Christine Mlot, "Tallying Nitrogen's Increasing Impact," *Science News* 151 (1997): 100; Hugh Ross, "Life in Extreme Environments," *Facts & Faith* 11, second quarter 1997, 6-7; Richard A. Kerr, "Cores Document Ancient Catastrophe," *Science* 275 (1997): 1265; Hugh Ross, "'How's the Weather?'—Not a Good Question on Mars," *Facts & Faith* 11, fourth quarter 1997, 2-3; Stephen Battersby, "Pathfinder Probes the Weather on Mars," *Nature* 388 (1997): 612; Ron Cowen, "Martian Rocks Offer a Windy Tale," *Science News* 152 (1997): 84; Hugh Ross, "Earth Design Update: The Cycles Connected to the Cycles," *Facts & Faith* 11, fourth quarter 1997, 3; Hugh Ross, "Earth Design Update: One Amazing Dynamo," *Facts & Faith* 11, fourth quarter 1997, 4; Peter Olson, "Probing Earth's Dynamo," *Nature* 389 (1997): 337; Weiji Kuang and Jeremy Bloxham, "An Earth-Like Numerical Dynamo Model," *Nature* 389 (1997): 371-74; Xiaodong Song and Paul G. Richards, "Seismological Evidence for Differential Rotation of the Earth's Inner Core," *Nature* 382 (1997): 221-24; Wei-jia Su, Adam M. Dziewonski, and Raymond Jeanloz, "Planet Within a Planet: Rotation of the Inner Core of the Earth," *Science* 274 (1996): 1883-87; Stephen H. Kirby, "Taking the Temperature of Slabs," *Nature* 403 (2000): 31-34; James Trefil, "When the Earth Froze," *Smithsonian* (December 1999), 28-30; Arnold L. Miller, "Biotic Transitions in Global Marine Diversity," *Science* 281 (1998): 1157-60; D. F. Williams et al., "Lake Baikal Record of Continental Climate Response to Orbital Insolation During the Past 5 Million Years," *Science* 278 (1997): 1114-17; S. C. Myneni, T. K. Tokunaga, and G. E. Brown Jr., "Abiotic Selenium Redox Transformations in the Presence of Fe(II, III) Oxides," *Science* 278 (1997): 1106-9; G. P. Zank and P. C. Frisch, "Consequences of a Change in the Galactic Environment of the Sun," *Astrophysical Journal* 518 (1999): 965-73; D. E. Trilling, R. H. Brown, and A. S. Rivkin, "Circumstellar Dust Disks Around Stars with Known Planetary Companions," *Astrophysical Journal* 529 (2000): 499-505; Joseph J. Mohr, Benjamin Mathiesen, and August E. Evrard, "Properties of the Intracluster Medium in an Ensemble of Nearby Galaxy Clusters," *Astrophysical Journal* 517 (1999): 627–49; Gregory W. Henry et al., "Photometric and Ca II and K Spectroscopic Variations in Nearby Sun-Like Stars with Planets," *Astrophysical Journal* 531 (2000): 415–37; Kimmo Innanen, Seppo Mikkola, and Paul Wiegert, "The Earth-Moon System and the Dynamical Stability of the Inner Solar System," *Astronomical Journal* 116 (1998): 2055–57; J. Q. Zheng and M. J. Valtonen, "On the Probability That a Comet That Has Escaped from Another Solar System Will Collide with the Earth," *Monthly Notices of the Royal Astronomical Society* 304 (1999): 579–82; Gregory Laughlin and Fred C. Adams, "The Modification of Planetary Orbits in Dense Open Clusters," *Astrophysical Journal Letters* 508 (1998): L171–L174; Shahid Naeem and Shibin Li, "Biodiversity Enhances Ecosystem Reliability," *Nature* 390 (1997): 507–9; S. H. Rhie et al., "On Planetary Companions to the MACHO 98-BLG-35 Microlens Star," *Astrophysical Journal* 533 (2000): 378–91; Daniel P. Schrag and Paul F. Hoffman, "Life, Geology, and Snowball Earth," *Nature* 409 (2001): 306; Craig R. Dina and Alexandra Navrotsky, "Possible Presence of High-Pressure Ice in Cold Subducting Slabs," *Nature* 408 (2000): 844–47; D. Vokrouhlicky and P. Farinella, "Efficient Delivery of Meteorites to the Earth from a Wide Range of Asteroid Parent Bodies," *Nature* 407 (2000): 606–8; Yumiko Watanabe, Jacques E. J. Matini, and Hiroshi Ohmoto, "Geochemical Evidence for Terrestrial Ecosystems 2.6 Billion Years Ago," *Nature* 408 (2000): 574–78; Hugh Ross, "Bacteria Help Prepare Earth for Life," *Connections* 3, first quarter 2001, 4; Crisogono Vasconcelos and Judith A. McKenzie, "Sulfate Reducers—Dominant Players in a Low-Oxygen World?" *Science* 290 (2000): 1711–12; Matthias Labrenz et al., "Formation of Sphalerite (ZnS) Deposits in Natural Biofilms of Sulfate-Reducing Bacteria," *Science* 290 (2000): 1744–47; Jochen Erbacher, Brian T. Huber, Richard D. Morris, and Molly Markey, "Increased Thermohaline Stratification as a Possible Cause for an Ocean Anoxic Event in the Cretaceous Period," *Nature* 409 (2001): 325–27; M. M. M. Kuypers, R. D. Pancost, J. S. A. Sinninghe Damsté, "A Large and Abrupt Fall in Atmospheric $CO_2$ Concentrations During Cretaceous Times," *Nature* 399 (1999): 342–45; Subir K. Banerjee, "When the Compass Stopped Reversing Its Poles," *Science* 291 (2001): 1714–15; Fred C. Adams and Gregory Laughlin, "Constraints on the Birth Aggregate of the Solar System," http://xxx.lanl.gov/abs/astro-ph/0011326, accessed September 6, 2001; Ian A. Bonnell, Kester W. Smith, Melvyn B. Davies, and Keith Horne, "Planetary Dynamics in Stellar Clusters," *Monthly Notices of the Royal Astronomical Society* 322 (2001): 859–65; Aylwyn Scally and Cathie Clarke, "Destruction of Protoplanetary Disks in the Orion Nebula," *Monthly Notices of the Royal Astronomical Society* 325 (2001): 449–56; Guillermo Gonzalez, Donald Brownlee, and Peter Ward, "The Galactic Habitable Zone: I. Galactic Chemical Evolution," *Icarus* 152 (2001): 185–200; Qingjuan Yu and Scott Tremaine, "Resonant Capture by Inward-Migrating Planets," *Astronomical Journal* 121 (2001): 1736–40; Zhang Peizchen, Peter Molnar, and William R. Downs, "Increased Sedimentation Rates and Grain Sizes 2-4 Myr Ago Due to the Influence of Climate Change on Erosion Rates," *Nature* 410 (2001): 891–97; N. Murray and M. Holman, "The Role of Chaotic Resonances in the Solar System," *Nature* 410 (2001): 773–79; O. Neron de Surgy and J. Laskar, "On the Long-Term Evolution of the Spin of the Earth," *Astronomy and Astrophysics* 318 (1997): 975–89; Richard A. Kerr, "An Orbital Confluence Leaves Its Mark," *Science* 292 (2001): 191; James C. Zachos et al., "Climate Response to Orbital Forcing Across the Oligocene-Miocene Boundary," *Science* 292 (2001): 274–78; John Bally and Bo Reipurth, "When Star Birth Meets Star Death: A Shocking Encounter," *Astrophysical Journal Letters* 552 (2001): L159–L162; Jon Copley, "The Story of O," *Nature* 410 (2001): 862–64; N. H. Sleep, K. Zahnle, and P. S. Neuhoff, "Initiation of Clement Conditions on the Earliest Earth," *Proceedings of the National Academy of Scienc-*

*es, USA* 98 (2001): 3666-72; Henry B. Throop et al., "Evidence for Dust Grain Growth in Young Circumstellar Disks," *Science* 292 (2001): 1686-89; G. Iraelean, N. C. Santos, M. Mayor, and R. Rebolo, "Evidence for Planet Engulfment by the Star HD82943," *Nature* 411 (2001): 163-66; M. Emilio, J. R. Kuhn, R. I. Bush, and P. Scherrer, "On the Constancy of the Solar Diameter," *Astrophysical Journal* 543 (2000): 1037-40; Q. R. Ahmad et al., "Measurement of the Rate of NUe + d > P+P+ e–Interactions Produced by 8B Solar Neutrinos at the Sudbury Neutrino Observatory," *Physical Review Letters* (2001): in press; Qingjuan Yu and Scott Tremaine, "Resonant Capture by Inward-Migrating Planets," *Astronomical Journal* 121 (2001): 1736-40; Chadwick A. Trujillo and David C. Jewitt, "Properties of the Trans-Neptunian Belt: Statistics from the Canada-France-Hawaii Telescope Survey," *Astronomical Journal* 122 (2001): 457-73; T. A. Michtchenko and S. Ferraz-Mello, "Resonant Structure of the Outer Solar System in the Neighborhood of the Planets," *Astronomical Journal* 122 (2001): 474-81.

## Appendix C: Fine-Tuning for Life in the Universe

*Appendix C is based on research found in many different resources. Rather than identify each factual source with numbers scattered throughout the appendix, the following are relevant sources arranged in approximately the order in which they relate to the material in the appendix:*

1. Hugh Ross, *The Fingerprint of God*, 2nd ed. (Orange, CA: Promise, 1991), 120-28; John D. Barrow and Frank J. Tipler, *The Anthropic Cosmological Principle* (New York: Oxford University Press, 1986), 123-57; Bernard J. Carr and Martin J. Rees, "The Anthropic Principle and the Structure of the Physical World," *Nature* 278 (1979): 605-12; John M. Templeton, "God Reveals Himself in the Astronomical and in the Infinitesimal," *Journal of the American Scientific Affiliation* 36 (1984): 194-200; Jim W. Neidhardt, "The Anthropic Principle: A Religious Response," *Journal of the American*

*Scientific Affiliation* 36 (1984): 201–7; Brandon Carter, "Large Number Coincidences and the Anthropic Principle in Cosmology," *Proceedings of the International Astronomical Union Symposium No. 63: Confrontation of Cosmological Theories with Observational Data*, ed. M. S. Longair (Boston: Reidel Publishing, 1974), 291–98; John D. Barrow, "The Lore of Large Numbers: Some Historical Background to the Anthropic Principle," *Quarterly Journal of the Royal Astronomical Society* 22 (1981): 404–20; Alan Lightman, "To the Dizzy Edge," *Science* 82 (1982): 24–25; Thomas O'Toole, "Will the Universe Die by Fire or Ice?" *Science* 81 (1981): 71–72; Fred Hoyle, *Galaxies, Nuclei, and Quasars* (New York: Harper & Row, 1965), 147–50; Bernard J. Carr, "On the Origin, Evolution, and Purpose of the Physical Universe," in *Physical Cosmology and Philosophy*, ed. John Leslie (New York: Macmillan, 1990), 134–53; Richard Swinburne, "Argument from the Fine-Tuning of the Universe," in *Physical Cosmology and Philosophy*, ed. John Leslie (New York: Macmillan, 1990), 154–73; R. E. Davies and R. H. Koch, "All the Observed Universe Has Contributed to Life," *Philosophical Transactions of the Royal Society of London*, B 334 (1991): 391–403; George F. R. Ellis, "The Anthropic Principle: Laws and Environments," in *The Anthropic Principle*, eds. F. Bertola and U. Curi (New York: Cambridge University Press, 1993), 27–32; Hubert Reeves, "Growth of Complexity in an Expanding Universe," in *Anthropic Principle*, 67–84; Oberhummer, Csótó, and Schlattl, "Stellar Production Rates of Carbon and Its Abundance in the Universe," *Science* 289 (2000): 88–90; Lawrence M. Krauss, "The End of the Age Problem and the Case for a Cosmological Constant Revisited," *Astrophysical Journal* 501 (1998): 461–66; Christopher C. Page et al., "Natural Engineering Principles of Electron Tunneling in Biological Oxidation-Reduction," *Nature* 402 (1999): 47–52; S. Perlmutter et al., "Measurements of $\Omega$ and $\Delta$ from 42 High-Redshift Supernovae," *Astrophysical Journal* 517 (1999): 565–86; P. deBarnardis et al., "A Flat Universe from High-Resolution Maps of the Cosmic Microwave Background Radiation," *Nature* 494 (2000): 955–59; A. Melchiorri et al., "A Measurement of Ω from the North American Test Flight of Boomerang," *Astrophysical Journal Letters* 536 (2000): L63–L66; Lawrence M. Krauss and Glenn D. Starkman, "Life, the Universe, and Nothing: Life and Death in an Ever-Expanding Universe," *Astrophysical Journal* 531 (2000): 22–30; Volker Bromm, Paolo S. Coppi, and Richard B. Larson, "Forming the First Stars in the Universe: The Fragmentation of Primordial Gas," *Astrophysical Journal Letters* 527 (1999): L5–L8; Jaume Garriga, Takahiro Tanaka, and Alexander Vilenkin, "Density Parameter and the Anthropic Principle," *Physical Review*, D 60 (1999): 5–21; Jaume Garriga and Alexander Vilenkin, "On Likely Values of the Cosmological Constant," *Physical Review*, D 61 (2000): 1462–71; Max Tegmark and Martin Rees, "Why Is the Cosmic Microwave Background Fluctuation Level $10^{-5}$?" *Astrophysical Journal* 499 (1998): 526–32; Jaume Garriga, Mario Livio, and Alexander Vilenkin, "Cosmological Constant and the Time of Its Dominance," *Physical Review*, D 61 (2000): in press; Peter G. van Dokkum et al., "A High Merger Fraction in the Rich Cluster MS 1054-03 at z = 0.83: Direct Evidence for Hierarchical Formation of Massive Galaxies," *Astrophysical Journal Letters* 520 (1999): L95–L98; Theodore P. Snow and Adolf N. Witt, "The Interstellar Carbon Budget and the Role of Carbon in Dust and Large Molecules," *Science* 270 (1995): 1455–57; Elliott H. Lieb, Michael Loss, and Jan Philip Solovej, "Stability of Matter in Magnetic Fields," *Physical Review Letters* 75 (1995): 985–89; B. Edvardsson et al., "The Chemical Evolution of the Galactic Disk. I. Analysis and Results," *Astronomy & Astrophysics* 275 (1993): 101–52; Hugh Ross, "Sparks in the Deep Freeze," *Facts & Faith* 11, first quarter 1997, 5–6; T. R. Gabella and T. Oka, "Detection of $H_3^+$ in Interstellar Space," *Nature* 384 (1996): 334–35; David Branch, "Density and Destiny," *Nature* 391 (1998): 23; Andrew Watson, "Case for Neutrino Mass Gathers Weight," *Science* 277 (1997): 30–31; Dennis Normile, "New Experiments Step Up Hunt for Neutrino Mass," *Science* 276 (1997): 1795; Joseph Silk, "Holistic Cosmology," *Science* 277 (1997): 644; Frank Wilczek, "The Standard Model Transcended," *Nature* 394 (1998): 13–15; Limin Wang et al., "Cosmic Concordance and Quintessence," *Astrophysical Journal* 530 (2000): 17–35; Robert Irion, "A Crushing End for our Galaxy," *Science* 287 (2000): 62–64; Roland Buser, "The Formation and Early Evolution of the Milky Way Galaxy," *Science* 287 (2000): 69–74; Joss Bland-Hawthorn and Ken Freeman, "The Baryon Halo of the Milky Way: A Fossil Record of Its Formation," *Science* 287 (2000): 79–83; Robert Irion, "Supernova Pumps Iron in Inside-Out Blast, *Science* 287 (2000): 203–5; Gary Gibbons, "Brane-Worlds," *Science* 287 (2000): 49–50; Anatoly Klypin, Andrey V. Kravtsov, and Octavio Valenzuela, "Where Are the Missing Galactic Satellites?" *Astrophysical Journal* 522 (1999): 82–92; Inma Dominguez et al., "Intermediate-Mass Stars: Updated Models," *Astrophysical Journal* 524 (1999): 226–41; J. Iglesias-Páramo and J. M. Vilchez, "On the Influence of the Environment in the Star Formation Rates of a Sample of Galaxies in Nearby Compact Groups," *Astrophysical Journal* 518 (1999) 94–102; Dennis Normile, "Weighing In on Neutrino Mass," *Science* 280 (1998): 1689–90; Eric Gawiser and Joseph Silk, "Extracting Primordial Density Fluctuations," *Science* 280 (1998): 1405–11; Joel Primack, "A Little Hot Dark Matter Matters," *Science* 280 (1998): 1398–1400; Stacy S. McGaugh and W. J. G. de Blok, "Testing the Dark Matter Hypothesis with Low Surface Brightness Galaxies and Other Evidence," *Astrophysical Journal* 499 (1998): 41–65; Nikos Prantzos and Joseph Silk, "Star Formation and Chemical Evolution in the Milky Way: Cosmological Implications," *Astrophysical Journal* 507 (1998): 229–40; P. Weiss, "Time Proves Not Reversible at Deepest Level," *Science News* 154 (1998): 277; E. Dwek et al., "The COBE Diffuse Infrared Background Experiment Search for the Cosmic Infrared Background. IV. Cosmological Implications," *Astrophysical Journal* 508 (1998): 106–22; G. J. Wasserburg and Y. Z. Qian, "A Model of Metallicity Evolution in the Early Universe," *Astrophysical Journal Letters* 538 (2000): L99–L102; Ron Cowen, "Cosmic Axis Begets Cosmic Controversy," *Science News* 151 (1997): 287; F. Gatti et al., "Detection of Environmental Fine Structure in the Low-Energy B-Decay Spectrum of 187 Re," *Nature* 397 (1997): 137–39; Ron Cowen, "Votes Cast for and Against the WIMP Factor," *Science News* 157 (2000): 155.

# BIBLIOGRAPHY

### General Sources on UFOs plan

Angelo, Joseph A. *The Extraterrestrial Encyclopedia: Our Search for Life in Outer Space.* New York: Facts on File, 1985.

Baker, Alan. *UFO Sightings.* New York: TV Books, 1999.

Baker, Robert A., and Joe Nickell. *Missing Pieces: How to Investigate Ghosts, UFOs, Psychics, and Other Mysteries.* Buffalo: Prometheus Books, 1992.

Blum, Howard. *Out There: The Government's Secret Quest for Extraterrestrials.* New York: Simon & Schuster, 1990.

Bryan, C. D. B. *Close Encounters of the Fourth Kind: Alien Abduction, UFOs, and the Conference at MIT.* New York: Knopf, 1995.

Bullard, Thomas E. *On Stolen Time.* Mount Rainier, MD: Fund for UFO Research, 1987.

———. *The Sympathetic Ear: Investigations as Variables in UFO Abduction Reports.* Mount Rainier, MD: Fund for UFO Research, 1995.

Catoe, Lynn E. *UFOs and Related Subjects: An Annotated Bibliography.* Detroit: Gale Research, 1978.

Clark, Jerome. *The UFO Encyclopedia,* 2nd ed., 2 vols. Detroit: Omnigraphics, 1998.

Crowe, Michael J. *The Extraterrestrial Life Debate, 1750–1900.* Mineola, NY: Dover Publications, 1999.

Davies, Paul. *Are We Alone?* New York: Basic Books, 1995.

Denton, Michael J. *Nature's Destiny.* New York: Free Press, 1998.

Devereux, Paul, and Peter Brookesmith. *UFOs and Ufology: The First Fifty Years.* New York: Facts on File, 1997.

Ellwood, Robert S., and Harry B. Partin. *Religious and Spiritual Groups in Modern America,* 2nd ed. Upper Saddle River, NJ: Prentice-Hall, 1988.

Fawcett, Lawrence, and Barry J. Greenwood. *The UFO Cover-Up: What the Government Won't Say.* New York: Prentice-Hall, 1984.

Fuller, John G. *The Interrupted Journey.* New York: Dial Press, 1966.
Hall, Richard H. *The UFO Evidence.* Lanham, MD: Scarecrow Press, 2001.
Hopkins, Budd. Intruders: *The Incredible Visitations at Copley Woods.* New York: Random House, 1987.
_____. *Missing Time: A Documented Study of UFO Abductions.* New York: Richard Marek, 1981.
Hynek, J. Allen. *The UFO Experience: A Scientific Inquiry.* New York: Marlowe, 1998.
Hynek, J. Allen, and Jacques Vallée. *The Edge of Reality: A Progress Report on Unidentified Flying Objects.* Chicago: Henry Regnery, 1975.
Jacobs, David M. *Secret Life: Firsthand Accounts of UFO Abductions.* New York: Simon & Schuster, 1992.
_____. *The UFO Controversy in America.* Bloomington: Indiana University Press, 1975.
Jung, Carl G. *Flying Saucers: The Myth of Things Seen in the Skies.* London: Kegan Paul, 1958.
Keel, John. *Disneyland of the Gods.* New York: Amok Press, 1988.
_____. *The Mothman Prophecies.* New York: Saturday Review Press, 1975.
_____. *UFOs: Operation Trojan Horse.* New York: Putnam, 1970.
Keyhoe, Donald E. *The Flying Saucers Are Real.* New York: Fawcett, 1950.
King, George. *The Twelve Blessings of Jesus.* Hollywood, CA: Aetherius Society, 1958.
Klass, Philip. *Bringing UFOs down to Earth.* Amherst, NY: Prometheus Books, 1997.
_____. *UFO Abductions: A Dangerous Game.* Buffalo, NY: Prometheus Books, 1988.
_____. *UFOs Explained.* New York: Random House, 1974.
_____. *UFOs: The Public Deceived.* Buffalo, NY: Prometheus Books, 1983.
Krauss, Lawrence M. *The Physics of Star Trek.* New York: Basic Books, 1995.
Lewis, James R., ed. *The Gods Have Landed.* New York: State Uni-

versity of New York Press, 1995.

Mack, John. *Abduction: Human Encounters with Aliens.* New York: Scribner, 1994.

Melton, J. Gordon. *Encyclopedia of American Religions*, 4th ed. Detroit: Gale Research, 1992.

Melton, J. Gordon, and George Eberhart. *The Flying Saucer Contactee Movement: 1950–1990.* Santa Barbara, CA: Santa Barbara Centre for Humanistic Studies, 1990.

Melton, J. Gordon, and Robert L. Moore. *The Cult Experience.* New York: Pilgrim, 1982.

Shapiro, Robert. *Origins: A Skeptic's Guide to the Creation of Life on Earth.* New York: Summit, 1986.

Spencer, John, ed. *The UFO Encyclopedia.* New York: Avon Books, 1991.

Sprinkle, R. Leo. *UFO Contactees and New Science.* Laramie, WY: School of Extended Studies, 1990.

Story, Ronald D., ed. *The Encyclopedia of UFOs.* Garden City, NY: Doubleday, 1980.

Strieber, Whitley. *Communion: A True Story.* New York: Beech Tree, 1987.

———. *Transformation: The Breakthrough.* New York: William Morrow, 1988.

*UFO: The Continuing Enigma.* Pleasantville, NY: Readers Digest Association, 1991.

Vallée, Jacques. *Confrontations: A Scientist's Search for Alien Contact.* New York: Ballantine Books, 1990.

———. *Dimensions: A Casebook of Alien Contact.* New York: Ballantine Books, 1988.

———. *The Invisible College: What a Group of Scientists Have Discovered About UFO Influences on the Human Race.* New York: Dutton, 1975.

———. *Messengers of Deception: UFO Contacts and Cults.* Berkeley, CA: And/Or Press, 1979.

———. *Revelations: Alien Contact and Human Deception.* New York: Ballantine Books, 1991.

Ward, Peter D., and Donald Brownlee. *Rare Earth.* New York: Copernicus, 2000.

Yockey, Hubert P. *Information Theory and Molecular Biology.* New York: Cambridge University Press, 1992.

**Christian Assessments of UFOs**

Alnor, William M. *UFOs in the New Age.* Grand Rapids, MI: Baker, 1992.
_____. *UFO Cults and the New Millennium.* Grand Rapids, MI: Baker, 1998.
Ankerberg, John, and John Weldon. *The Facts on UFOs and Other Supernatural Phenomena.* Eugene, OR: Harvest House, 1992.
Enroth, Ronald. *The Lure of the Cults.* Downers Grove, IL: InterVarsity, 1987.
Hymers, R. L. *Encounters of the Fourth Kind.* Van Nuys, CA: Bible Voice, 1976.
Martin, Walter R. *Chariots of the Who?* Walter Martin's Religious InfoNet. Audiocassette.
_____. *UFOs: Friend, Foe or Fantasy?* Walter Martin's Religious InfoNet. Audiocassette.
Rhodes, Ron. *Alien Obsession.* Eugene, OR: Harvest House, 1998.
Ross, Hugh. *Beyond the Cosmos,* 2nd ed. Colorado Springs, CO: NavPress, 1999.
_____. *The Creator and the Cosmos,* 3rd ed. Colorado Springs, CO: NavPress, 1999.
_____. *UFOs...The Mystery Resolved.* Reasons To Believe. Videocassette.
Weldon, John, and Zola Levitt. *Encounters with UFOs.* Irvine, CA: Harvest House, 1975.
Wilson, Clifford, and John Weldon. *Close Encounters: A Better Explanation.* San Diego: Master, 1978.

**General Sources on the New Age Movement**

Melton, J. Gordon. *New Age Encyclopedia.* Detroit: Gale Research, 1990.
Melton, J. Gordon, Jerome Clark, and Aidan A. Kelly. *New Age Almanac.* New York: Visible Ink, 1991.

## Christian Critiques of the Occult, Demonism, and the New Age

Arnold, Clinton E. *Powers of Darkness: Principalities and Powers in Paul's Letters.* Downers Grove, IL: InterVarsity, 1992.

Barnhouse, Donald Grey. *The Invisible War.* Grand Rapids, MI: Zondervan, 1965.

Bromiley, Geoffrey W., gen. ed. *International Standard Bible Encyclopedia.* 4 vols. Grand Rapids, MI: Eerdmans, 1979. See biblical terms relating to occultic practices.

Clark, David K., and Norman L. Geisler. *Apologetics in the New Age: A Christian Critique of Pantheism.* Grand Rapids, MI: Baker, 1990.

Elwell, Walter A., ed. *Evangelical Dictionary of Biblical Theology.* Grand Rapids, Mich.: Baker, 1996. See biblical terms relating to occultic practices.

Geisler, Norman L., and J. Yutaka Amano. *The Infiltration of the New Age Movement.* Wheaton, IL: Tyndale, 1989.

Green, Michael. *Exposing the Prince of Darkness.* Ann Arbor, MI: Servant, 1991.

Groothuis, Douglas R. *Confronting the New Age.* Downers Grove, IL: InterVarsity, 1988.

\_\_\_\_\_. *Revealing the New Age Jesus.* Downers Grove, IL: InterVarsity, 1990.

\_\_\_\_\_. *Unmasking the New Age Movement.* Downers Grove, IL: InterVarsity, 1986.

Hawkins, Craig. *Goddess Worship, Witchcraft, and Neo-Paganism.* Grand Rapids, MI: Zondervan, 1998.

\_\_\_\_\_. *Witchcraft: Exploring the World of Wicca.* Grand Rapids, MI: Baker, 1996.

Hoyt, Karen, ed. *The New Age Rage.* Old Tappan, NJ: Revell, 1987.

Koch, Kurt. *Christian Counseling and Occultism.* Grand Rapids, MI: Kregel, 1972.

\_\_\_\_\_. *Demonology: Past and Present.* Grand Rapids, MI: Kregel, 1973.

\_\_\_\_\_. *The Devil's Alphabet.* Grand Rapids, MI: Kregel, 1984.

Kole, Andre, and Terry Holley. *Astrology and Psychic Phenomena.* Grand Rapids, MI: Zondervan, 1998.

Lewis, C. S. *The Screwtape Letters*. New York: Macmillan, 1961.
Martin, Walter R. *Screwtape Writes Again*. Santa Ana, CA: Vision House, 1975.
Miller, Elliot. *A Crash Course on the New Age Movement*. Grand Rapids, MI: Baker, 1989.
Montgomery, John Warwick. *Demon Possession*. Minneapolis: Bethany, 1976.
_____. *Principalities and Powers*. Minneapolis: Dimension Books, 1975.
Newport, John P. *The New Age Movement and the Biblical Worldview*. Grand Rapids, MI: Eerdmans, 1998.
North, Gary. *Unholy Spirits*. Tyler, TX: Institute for Christian Economics, 1994.
Passantino, Bob and Gretchen. *Satanism*. Grand Rapids, MI: Zondervan, 1995.
Rhodes, Ron. *The Counterfeit Christ of the New Age Movement*. Grand Rapids, MI: Baker, 1990.
_____. *New Age Movement*. Grand Rapids, MI: Zondervan, 1995.
Wilson, Clifford, and John Weldon. *Occult Shock and Psychic Forces*. San Diego: Master, 1980.

# Name Index

47 Tucanae, 40–41
51 Pegasi, 39

Abduction Study Conference, 138
Adamski, George, 152–153, 155
Aetherius Society, 162–163, 178, 236n3, 238n7
Air Force (U.S.), 15–16, 24–25, 72, 79, 81, 83–90, 104, 224n8
Amalgamated Flying Saucer Clubs of America, 236n3
American Astronomical Society, 127
Andrews Air Force Base, 85
Applewhite, Marshall Herff, 165–167
Aquinas, Thomas, 174
Area 51, 10, 79–80, 88–90, 93, 104
Arnold, Kenneth, 14–15, 80, 84
Ascended Masters, 151, 163, 185
Ashtar Command, 236n3
Association of Sananda and Sanat Kumara, 236n3
Aurora, 89

Bada, Jeffrey, 45
Baker, Alan, 72
Baker, Robert A., 145
Balch, Robert W., 165–167
Ballard, Edna, 151
Ballard, Guy, 151
Blavatsky, Madame Helena Petrovna, 151
Bounias, Michel, 73
Brazel, W. W. "Mac," 80, 84
Brotherhood of the Seven Rays, 160
BUFORA (British UFO Research Association), 20
Bullard, Thomas E., 135–136, 138–140, 142, 144, 232n1

Cabell, Charles P., 85

Carter, Jimmy, 77
Catoe, Lynn, 132
Central Intelligence Agency (CIA), 85–86, 100
Chyba, Christopher, 49
Clark, Jerome, 15, 137
Cold War, 15, 82, 90, 92, 98–99
Condon, Edward U., 86–87
Condon Report, 15–16, 80, 86–87
Cosmic Circle of Fellowship, 236n3
Cosmic Star Temple, 236n3
CUFOS (Center for UFO Studies), 20, 25, 28, 30, 72, 87, 225n24

Davies, Paul, 132
Dawkins, Richard, 48
Deval UFO, Inc., 236n3
Dicke, Robert, 42

Earth Mission Interplanetary Outreach, 160
Eberhart, George M., 153
Einstein, Albert, 106–107
Ellis, George, 106
Ellwood, Robert S., 156, 161–162
Elohim, 168
Europe, 69, 92, 225n16

Federation of American Scientists, 89
Foo Fighters, 14
Foster, J. B. sheep ranch, 80–81
Freedom of Information Act (FOIA), 98
Friedman, Stanton, 33
Fuller, John G., 137

Gallup Organization, 23, 100
Ganymede, 45

## Name Index

George Adamski Foundation, 236n3
Good, Timothy, 24
Groom Lake, 79–80, 88–90

Hale Bopp comet, 16, 167
Haut, Walter, 81, 83
Hawking, Stephen, 106, 176
Hynek, J. Allen, 20, 24–27, 72, 86–87, 129, 132
Heaven's Gate, 16, 165–167, 185
Hill, Barney, 136–137
Hill, Betty, 136–137
Hitler, Adolph, 92–93
Hopkins, Budd, 137–138, 146
Horgan, John, 58
Hoyle, Fred, 53
Hubble Space Telescope, 40–41, 64
Human Individual Metamorphosis (HIM), 166

I AM movement, 151
Internal Communications Research Services of Media, 24
Internal Society for the Study of the Origin of Life, 57
Interplanetary Parliament, 162
Iran-Contra hearings, 100

Jacobs, David M., 146
Jastrow, Robert, 105
Jervis, Robert, 95–97
Jesus Christ, 117–118, 120, 131, 152, 162–164, 178–180, 238n7
Jesusonian Foundation (Urantia), 236n3
Jung, Carl G., 32
Jupiter, 39, 44–45, 70, 126, 198, 205

Keel, John A., 129, 132, 146, 156–157
Khomeini, Ayatollah, 92
King, George, 162–163

Klass, Philip J., 32, 145, 224n13
Korean War, 85

Laibow, Rima E., 145
Last Day Messengers, 236n3
Lemaître, Jacques, 132
Lenin, Vladimir, 92–93, 102
Lunar and Planetary Laboratory, 53
Lunar and Planetary Science Conference, 53

Mack, John E., 146
Marcel, Jesse, 81, 83
Mark-Age Inc., 236n3
Mars, 52, 126, 152, 156, 164
Marx, Karl, 101–102
McCampbell, James, 132
Melosh, Jay, 53
Melton, J. Gordon, 150–151, 153, 163–164
Menzel, Donald, 32
Military (U. S.), 8, 10, 14–15, 19, 24, 29, 83, 87–90, 98, 100, 102–103, 224n4
Milky Way Galaxy, 40–41, 46, 64, 187–188
Miller, Elliot, 235n12
Moore, Charles, 82–83
Morowitz, Harold, 54–55, 58
MUFON (Mutual UFO Network), 20
Murchison meteorite, 53

Nash, Ronald H., 169
National Academy of Sciences, 87
National Aeronautics and Space Administration (NASA), 18–19, 32, 62, 64, 77, 126
Nellis Air Force Base, 88
Nettles, Bonnie Lu, 165–167
New Age, 20, 128, 131, 154, 162, 164, 166, 178, 235n12
New Thought, 162

Newton, Irving, 81
New York University, 81, 83
Norman, Ernest L., 163–164
Norman, Ruth E., 163–165
Oberg, James, 32
Ockham, William of, 102
Ockham's Razor, 93, 102–103, 170
One World Family, 236n3
Orgel, Leslie, 57

Palmer, Susan Jean, 237n24
Partin, Harry B., 156, 161–162
Penrose, Roger, 106
People's Republic of China (PRC), 90
Persinger, Michael A., 32, 145
Project Blue Book, 10, 15, 24–25, 72, 79–80, 84–88, 90, 93, 104
Project Grudge, 15, 24, 84–85, 87
Project Mogul, 81–83, 224n8
Project Sign, 15, 24, 84, 87
PSR 1257+12, 39

Raelian Movement, 167–169, 236n3
Ramey, Roger, 81, 83–84
Reverend Kirk, 126
Rhodes, Ron, 177
Robertson, H. P., 85
Robertson Panel, 85
Roswell, 10, 79–84, 90, 93, 104, 224n4

Sagan, Carl, 32, 38, 41, 49, 145, 176
Saliba, John A., 25, 31, 156, 161
Saturn, 39, 126, 153
Seamans, Robert C., 87
Semjase Silver Star Center, 236n3
SETI, 19, 63
Shapiro, Robert, 53, 57
Smith, Walter Bedell, 85

Solar Cross Foundation, 236n3
Solar Light Retreat, 236n3
Spencer, John, 27
SR-71 Blackbird program, 29, 86, 88
Stalin, Joseph, 92–93
Star Light Fellowship, 160
Stevenson, David, 45
Strieber, Whitley, 128–129, 145–146
Sturrock, Peter, 127–128
Swedenborg, Emanuel, 150–151
Swendenborgianism, 150–151

Tantra Buddhism, 162
Theosophical Society, 151, 165
Total Overcomer's Anonymous, 167
Truman Doctrine, 85
Truman, Harry, 224n4

Unarius, 163–165, 236n3
United States, 14–15, 19, 24, 85, 88, 92–94, 160, 169
Universariun Foundation, 236n3
Universe Society Church Science of Life (UNISOC), 236n3
University of Colorado UFO project, 86, 88
Urantia Book, The, 130–131, 151–152, 178, 238n7

Vallée, Jacques, 24–27, 72, 115, 123, 131, 139, 146
Van Tassel, George, 154
Venus, 8, 28, 126, 152–153, 156, 162, 164, 183
Vietnam War, 100
Vorilhon, Claude, 167–169

Walton, Travis, 137
Watergate, 100
Weldon, John, 146
White Star, 236n3
Whitmore, John, 140–141
Wills, Christopher, 45
World Understanding, 236n3
World War II, 14, 99, 125

# Subject Index

abductees, 135–147, 150, 155–156, 185
abduction, 10, 16, 20, 74, 135–147, 149–150
agnostic, 131–132, 176
amino acids, 41, 50, 52–56
angel, 35, 119–121, 173–174, 178
angel hair, 75
animals, 48, 73–74, 129–131, 155, 195, 200, 207
anomaly, 26–27
anthropic principle, 111–112
apologists, 8, 21, 146, 177, 235n12
asteroids, 44–45, 49, 52–53, 62–63, 194, 198, 205
astrology, 133, 165
astronomy, 8, 13, 38–46, 63, 108, 126–127
atmosphere, 9, 34, 45, 51, 153, 192–196, 198–199, 200, 206
atmospheric physics, 192–195, 198–200, 205–206
atmospheric turbulence, 28
autokinesis, 28
automatic writing, 20, 151, 155, 177
aurora borealis, 29
aviation, 14, 29, 98, 154

big bang, 105–107, 210
biospheres, 57
black hole, 66–67
boron, 41, 206

carbon, 41–42, 49, 51, 68, 195–196, 204, 206–207, 209
carbonate-silicate cycle, 45, 197
catalyst, 56
cell, 49, 54, 57–58, 111
cell membrane, 57–58
channeling, 20, 150, 152, 154–155, 160–165, 177–178, 185, 235n12

Christian, 9, 21, 35, 120, 130, 133, 146, 166, 173–181, 177–179
Christology, 178
close encounters of the first kind (CE-1), 25–27, 125, 130, 133
    second kind (CE-2), 26, 130, 133
    third kind (CE-3), 25–26, 133
    fourth kind (CE-4), 25–26, 74, 130, 133, 135
    fifth kind (CE-5), 25–26, 133
comets, 16, 28, 44–46, 49–50, 52–53, 62–64, 165, 167, 184, 191, 194, 198, 201, 204–205, 207
cosmic inflation models, 106–107
cosmos, 14, 38–39, 41, 48, 53–54, 58–59, 61, 106–108, 111–112, 115–119, 175–176
conspiracy, 10, 19, 80, 87, 100–101, 114, 184
conspiracy theory, 80–82, 84, 90–95, 97–104, 224n13, 225n24
contactees, 20, 132, 147, 149–157, 161–162, 177–178, 185
Creator, 108–109, 112, 116–118, 175–176, 178, 184
cryptogamic colonies, 45
crystals, 20, 28, 86
cult, 10, 16, 20–21, 128, 130–131, 150–152, 157, 159–171, 178, 185
cytosine, 57

data-driven thinking, 94–95
Daylight Disks (DD), 26
deists, 176
deleterious mutations, 68–69
dematerialization, 177
demon, 119–121, 131–133, 146, 174, 177, 179–180, 185–186
demonic, 21, 35, 132–133, 146, 179–180
demonology, 132
design, 111, 118
divination, 177
DNA, 52–53, 56–57, 168
doctrine of transcendence, 108, 116–119

Earth, 8, 34, 39, 44–46, 48–49, 50–55, 57–58, 62–63, 70, 110–111, 118–120, 126, 156, 184–185, 187–201, 203–207, 210

## Subject Index 271

Eastern mystical, 154, 162, 164
electromagnetic force, 110, 209
electromagnetic signals, 76
electrons, 65, 209
energy, 64, 67, 106–108, 117, 119, 184, 196, 209
enzyme, 56
ESP, 142
evolution, 19–20, 47–59, 106, 151, 156, 167, 178, 184
exobiology, 19
exotheology, 21
exotic chemistry, 41, 44–46
extinction, 65, 68–69
extrasolar planets, 39–40, 44c
extraterrestrial, 8, 14, 16, 18–21, 31, 34–35, 39, 41, 46, 50, 57, 62, 85, 87, 92, 99, 146, 150–154, 160, 176
extraterrestrial hypothesis (ETH), 19, 33–34, 38, 48, 58–59, 70, 114, 174–176, 181, 183
extraterrestrial intelligence, 9–11, 14, 18, 21, 38, 41, 174–175, 183–184

fallen angels, 119–121
flyby, 26–27
flying saucer, 14–15, 139–140, 153, 160, 166–167
"flying saucer age," 14–17, 24, 150, 183
flying saucer conventions, 203
fossil record, 49, 55

galaxy, 40–42, 46, 63–64, 108–109, 111–112, 187–188, 190, 203, 209
gas giant, 39–40, 63, 205
general relativity, 65–67, 106–107, 114–115
geocentric, 102
globular cluster, 40–41
government cover-ups, 10, 16, 19, 33, 79–90, 103, 114, 184
gravitational force, 110, 209
gravity, 34, 45, 62–63, 66, 114, 188, 192, 205

"Grays, The" 143, 155
hallucinatory experience, 31–32, 144–145, 179
healing, 20, 140, 164
heliocentric, 102
helium, 31, 40, 111, 209
hoax, 15, 20, 30–31, 72–73, 131, 137, 145, 153, 176, 183
humanoid, 130, 140, 143, 155, 168, 224n4
hybrid race, 144, 146
hydrocarbon, 47, 50, 195
hydrogen, 40, 45, 50
hypnosis, 82, 135, 141–142, 145
hypnotic, 141–142, 144

IFO (identified flying object), 27–32, 72, 119, 174
inter-dimensional beings (interdimensional beings), 20–21
interdimensional hypothesis (IDH), 19, 33–35, 112–121, 183–184
intergalactic travel, 34, 61–70, 114, 184
interplanetary, 33, 52, 63, 154, 162
interplanetary media, 44
interstellar, 16, 34, 40, 53–54, 62, 114, 204
isotopes, 51

laws of physics, 34, 62, 64, 67–68, 71, 75, 125, 131, 175, 184
levitation, 135, 177
life chemistry, 40–43, 45, 48–59, 111, 187–188, 191
life sites, 40, 43–46, 111, 114
life span, 58, 68
lunar meteorites, 49

magnetic field, 45, 193, 198, 204–205
magnetosphere, 45, 193
maneuver, 26–27, 63
Martian, 52, 153
mass, 39, 42, 45, 63–66, 106–107, 188, 190, 194, 203–205, 209–210
matter, 54, 64, 66, 76, 106–108, 110, 117, 119, 184, 203, 210

mediumship, 151–152, 162, 164, 177, 235n13
metabolism, 43, 45
metal rich stars, 40
metaphysical, 20, 152, 156, 162, 166, 178, 235n13
meteorites, 50, 52–53, 64
monistic, 178
mystical, 33, 154, 162, 164, 178

natural-explanation hypothesis, 18–19, 27–33, 58, 114, 183
naturalistic worldview, 8, 38, 48–49, 51–54, 57–58, 130
natural scientists, 18–19
near-death experiences (NDEs), 145
neurological stimuli, 31
neutron, 43, 64–65, 209–210
neutron star, 39
Nocturnal Lights (NL), 26
"Nordics, The," 143, 155
nova, 63, 174
nuclei, 65

occult, 16, 34–35, 128, 131, 133, 146, 150–151, 153–154, 159, 162, 169, 177, 180, 185
occultism, 20, 150–151, 152, 165, 185
orbit, 39–40, 44–46, 62–63, 163, 187–190, 192, 194, 198, 203–205
organism, 48, 51, 55, 110, 184
origin-of-life, 43, 46–51, 56–57
out-of-body experiences (OBEs), 140–141
out-of-body travel, 20, 140–141
oxygen-ultraviolet paradox, 51–52, 191, 195, 199–200

paganism, 177
panspermia, 53–54, 58
pantheistic, 116
paranormal, 16, 35, 113, 126, 141–142, 145–146, 155, 157, 161, 177, 235n13

particles, 45, 50, 53, 65–67
photons, 65
planet, 8–9, 28, 37–49, 51–53, 58–59, 61–63, 69–70, 110–112, 150–151, 156, 164, 174–176, 183, 188–191, 193, 198, 203–205, 207–208
plate tectonic, 45, 189, 194, 196, 207
pluralism, 108
pluralistic, 141, 178
PNA, 53
poltergeists, 16, 20, 27, 132, 142
postbiotic molecules, 50–51
prebiotic molecules, 50–53
prebiotic soup, 51, 56–57
primordial soup, 50–51, 56
prophetic, 161
prophets, 168
protons, 43, 65, 191–193, 204, 209
psychic, 20, 25, 34, 132–133, 142–143, 145–146, 150–153, 155–156, 160–161, 166, 177
psychokinesis, 177
psychological, 17, 20, 28, 31–32, 85, 93–95, 114, 128–131, 141, 145, 179
psychological stimuli, 31

quantum physics, 114

radar-visual (RV), 26
radiation, 26, 39, 43–45, 53, 61, 63–65, 68, 73, 184, 187–188, 190–191, 193, 195–196, 199, 209
radioactive decay, 45
reincarnation, 164, 166–167, 178
religion (UFO), 20, 160–161, 165–167, 178, 185
residual UFOs (RUFOs), 32, 71–77, 114–115, 119–121, 123–133, 173–174, 179, 181, 183–185
RNA, 52–53, 55–58

satellite, 29, 44–46, 72

scientific method, 38, 58, 92, 114, 116, 174
Scripture, 9, 11, 116–120, 133, 147, 166, 173–175, 177–181, 185
sexual abuse, 74, 139–140, 145
sightings, 10, 14–16, 20, 24–27, 31–32, 70, 73, 76, 79–80, 84–88, 124–125, 142, 183
silicon, 41, 190, 207
"Skunk Works," 88
social scientists, 20, 101, 181
sociological stimuli, 31
solar nebula, 40, 189, 203
solar system, 20, 39–41, 51, 53, 62, 64, 70, 111–112, 118, 126, 150, 155–156, 162, 175, 190, 198, 204
spacecraft, 10, 16, 26, 33–34, 59, 62, 64–66, 69, 75, 79, 103, 137, 139–140, 147, 154–155, 160, 164, 166–167, 185
space-time dimensions, 66–67, 107, 114–116, 119, 131, 133, 156, 174, 230n2
space-time theorems, 106–107, 115, 184
spiritism, 20, 177
spiritualism, 133, 161, 164
stars, 28, 38–42, 44, 53, 62–63, 108, 110–111, 126, 176, 187–190, 204, 209
star ashes, 40
star birth, 40, 189
star death, 40
star formation, 40, 187–189, 204
supergiant star, 39, 53, 63
supernova, 39–40
supernova eruptions, 63, 188–189, 203, 210
supradimensional, 118, 121, 181

tectonic-geological phenomena, 29
technology, 29, 50, 61, 68–69, 90, 97, 99, 124–125
telepathy, 140–141, 154, 161–162, 177–178
teleportation, 177
theologian, 21, 125, 102, 132, 146, 177
theosophy, 151, 161–163

theory-driven thinking, 93–98
thermals, 28
time-space continuum, 34
trance, 20, 151, 155, 161, 163
trance-channeling, 20, 155, 177, 185
trans-dimensional, 10, 116–119, 186

ufologist, 18–19, 23–24, 26–27, 31, 33, 35, 71, 135, 137, 139, 143, 146, 149, 183
ufology, 14–15, 17–21, 23–27, 32, 137, 139, 146, 153
ultraterrestrials, 157
universe, 9, 14, 21, 38, 41, 43, 46, 54, 58, 67, 70, 77, 106–112, 114–117, 130–131, 142, 155, 175–176, 184–185, 207–210
universalistic, 178
unmanned aerial vehicles (UAVs), 90
U.S. government, 10, 15–16, 18–19, 33, 77, 79–94, 97–104, 184

vascular plants, 45, 207
velocity of light, 62, 64–65, 209
volcanic eruptions, 44, 194, 196, 207

worldview, 8–9, 160, 165–166, 169–170, 174–175, 178, 185
wormholes, 65–67, 184

# About the Authors

**Hugh Ross** is senior scholar and founder of Reasons to Believe (RTB). With a degree in physics from the University of British Columbia and a PhD in astronomy from the University of Toronto, Hugh—initially a skeptic, always curious, and eventually a Christian—continued his research on quasars and galaxies as a postdoctoral fellow at the California Institute of Technology. After five years there, he transitioned to full-time ministry. In addition to founding and leading RTB, he remains on the pastoral staff at Christ Church Sierra Madre. His writings include journal and magazine articles, blogs, and numerous books—*The Creator and the Cosmos, Why the Universe Is the Way It Is,* and *Designed to the Core,* among others. He has spoken on hundreds of university campuses as well as at conferences and churches around the world. Hugh lives in Southern California with his wife, Kathy.

**Kenneth Richard Samples** serves as senior research scholar at Reasons to Believe (RTB). Kenneth holds a BA in social science with an emphasis in history and philosophy from Concordia University and an MA in theological studies from Talbot School of Theology. Prior to joining RTB, he worked as senior research consultant and correspondence editor at the Christian Research Institute and regularly cohosted popular call-in radio program *The Bible Answer Man*. Kenneth's other books include *Christianity Cross-Examined, Classic Christian Thinkers,* and *God among Sages*. He leads RTB's *Clear Thinking* podcast and writes *Reflections,* a weekly blog dedicated to exploring the Christian worldview. Additionally, he is currently an adjunct instructor of apologetics at Biola University and has spoken at universities and churches around the world. Kenneth lives in Southern California with his wife, Joan. They have three grown children.

**Mark T. Clark** is professor emeritus of political science and director of the National Security Studies program at California State University, San Bernardino (CSUSB). He has published in a variety of scholarly venues on national security, nuclear weapons, arms control and intelligence, and the Christian perspective of just war doctrine. With the help of several grants, he has helped establish CSUSB as an Intelligence Community Center of Academic Excellence (ICCAE) and he codeveloped a new MS degree in National Cyber Security Studies (along with the MA in National Security Studies), the first of its kind in the nation. With George Mason University, he has worked as lead analyst to demonstrate how evidence-based reasoning using a learning agent, called Cogent, can improve critical and analytical thinking skills and improve intelligence analysis with such tools in a crowd-sourced environment. He also has led seven different open-source research projects for the Institute for Analysis of the National Security Agency (NSA). Mark served in the U.S. Marine Corps from 1973 to 1977. Mark and his wife, Mara, serve in their local church, Sierra Vista Community Church, in Upland, CA. They live in La Verne, California, and Durango, Colorado.

# About Reasons to Believe

Reasons to Believe (RTB) exists to reveal God in science. Based in Covina, California, RTB was established in 1986 and since then has taken scientific evidence for the God of the Bible across the US and around the world. Our ongoing work is providing content for all who desire to explore the connection between science and the Christian faith.

RTB is unique in its range of resources. The curious can explore articles, podcasts, and videos. Those who want to learn more can delve into books, in-person and livestreamed events, and online courses. Donors enable us to continue this important work.

For more information, visit reasons.org.

For inquiries, contact us via:
818 S. Oak Park Rd.
Covina, CA 91724
(855) REASONS | (855) 732-7667
ministrycare@reasons.org